SpringerBriefs in Electrical and Computer Engineering

Control, Automation and Robotics

Series editors

Tamer Başar
Antonio Bicchi
Miroslav Krstic

More information about this series at http://www.springer.com/series/10198

Ionela Prodan · Florin Stoican
Sorin Olaru · Silviu-Iulian Niculescu

Mixed-Integer Representations in Control Design

Mathematical Foundations and Applications

 Springer

Ionela Prodan
Laboratory of Conception and Integration
    of Systems
Université Grenoble Alpes
Valence
France

Florin Stoican
Department of Automatic Control
    and Systems Engineering
Politehnica University of Bucharest
Bucharest
Romania

Sorin Olaru
Laboratory of Signals and Systems
CentraleSupélec - CNRS - Université
    Paris-Sud, Université Paris-Saclay
Gif-sur-Yvette
France

Silviu-Iulian Niculescu
Laboratory of Signals and Systems
CNRS - CentraleSupélec - Université
    Paris-Sud, Université Paris-Saclay
Gif-sur-Yvette
France

ISSN 2191-8112   ISSN 2191-8120   (electronic)
SpringerBriefs in Electrical and Computer Engineering
ISSN 2192-6786   ISSN 2192-6794   (electronic)
SpringerBriefs in Control, Automation and Robotics
ISBN 978-3-319-26993-1   ISBN 978-3-319-26995-5   (eBook)
DOI 10.1007/978-3-319-26995-5

Library of Congress Control Number: 2015955893

Printed on acid-free paper

This Springer imprint is published by SpringerNature
The registered company is Springer International Publishing AG Switzerland

Preface

While the scope of a book is to present the problem discussed as esoterically as possible we also have to acknowledge its roots in "real life." This is the case here with "control problems with contradictory conditions" which are readily found in everyday life. They may not be stated so, but each time we balance between mutually exclusive goals or have to choose a direction in the detriment of another we actually solve (optimally or not) such a control problem. An "either/or" decision is always difficult for human beings, the same goes for optimization algorithms with discrete variables. We have thus strong reasons to pursue solutions for this class of problems. As is usually the case, the solution is better or faster if we understand the problem's underlying structure. Therefore, the scope of this book is to provide efficient constructions which are subsequently put under a mixed-integer form which can be solved by a computer in a "reasonable" time.

This book represents the culmination of over 5 years of collaboration work of the authors. The main contributions are the result of work started during the Ph.D. theses of the first two authors and by subsequent advancements in the following years.

This book was inspired by our desire to bring to light the importance of the analysis and control of dynamical systems with conflicting objectives and the effective usage of the associated mixed-integer formulations. It is worth mentioning that the topic is not new and monographs exists covering mixed-integer optimization. However, most of them assume a specific background in mathematics and optimization. The present book is dedicated to a generic class of constraints and their use in optimization problems and, in this respect, goes deeper into the details of their construction, representation and computational complexity. It is important to mention that the present manuscript is mainly dedicated to the problem description and not to the numerical optimization routines. It focuses on the mixed-integer aspects of the constraints formulation and their relationship with the optimization-based control design. To our knowledge, another textbook is not currently available that covers a compact treatment of the non-convex feasible set representation via mixed-integer representations, gathering the recent research advancements in the literature and illustrating the potential impact on optimization-based design, as for example in control design.

One of the most important features of the book is that it provides all along the manuscript the tools for easy reconstruction of the illustrative examples. The applications encompass important issues from control theory, ranging from motion planning with obstacle and collision avoidance and up to fault tolerant control schemes.

The book will hopefully not only serve the purpose of disseminating research results but also of raising the awareness for these challenging, timely and relevant research topics on optimization and control design. Moreover, we hope that this book will find attention in the diverse control engineering, computational mathematics and optimization communities and thus will contribute to the development of mixed-integer representations as a well-defined research field.

Valence   Ionela Prodan
Bucharest   Florin Stoican
Gif-sur-Yvette   Sorin Olaru
Gif-sur-Yvette   Silviu-Iulian Niculescu
September 2015

Acknowledgment The work of Florin Stoican is funded by the Sectorial Operational Programme Human Resources Development 2007–2013 of the Ministry of European Funds through the Financial Agreement [grant number POSDRU/159/1.5/S/132395].

Contents

Abbreviations

FDI	Fault Detection and Isolation
FTC	Fault Tolerant Control
KKT	Karas–Kuhn–Tucker
LMI	Linear Matrix Inequalities
LP	Linear Programming
LQ	Linear Quadratic
LTI	Linear Time Invariant
MILCP	Mixed-Integer Linear Constrained Programming
MILP	Mixed-Integer Linear Programming
MINLP	Mixed-Integer Non-Linear Programming
MIP	Mixed-Integer Programming
MIQCP	Mixed-Integer Quadratically Constrained Programming
MIQP	Mixed-Integer Quadratic Programming
MPC	Model Predictive Control
NP-hard	Non-deterministic Polynomial-time hard
PWA	Piecewise Affine
QP	Quadratic Programming
RC	Reconfiguration Control
RPI	Robust Positive Invariance
UAV	Unmanned Aerial Vehicle

Notation

The conventions and the notations used in the book are the ones classically employed in control theory literature.

- $\mathbb{R}$, $\mathbb{Z}$ and $\mathbb{N}$ denote the set of real numbers, the set of integers and the set of non-negative integers, respectively.
- $\mathbb{R}^n$ and $\mathbb{R}^{m \times n}$ denote the vector field with n elements and the real matrices field with n rows and m columns, respectively. The same notation is adopted for the sets of integer and non-negative integers.
- I denotes the identity matrix of appropriate dimension and e_i ($e_i^\top$) denotes its ith column (row).
- $\mathbf{1}$ denotes the matrix of ones and $\mathbf{0}$ the matrix of zeros of appropriate dimension.
- For a matrix $A \in \mathbb{R}^{n \times n}$, the standard notation $A^\top$ denotes the transpose of matrix A, A^{-1} denotes the inverse of matrix A and $A \succ 0$ ($A \succeq 0$) denotes a (strictly) positive definite matrix.
- $\|z\|_M$ is the weighted Euclidean norm, i.e., $\sqrt{(z^\top M z)}$.
- For a discrete-time signal $x \in \mathbb{R}^n$, the current and successor states are denoted as $x(k)$ and $x(k+1)$, respectively.
- Absolute values and vector inequalities are considered elementwise (unless otherwise explicitly stated), that is, $|T|$ denotes the elementwise magnitude of a matrix T and $x \leq y$ ($x < y$) denotes the set of elementwise (strict) inequalities between the components of the real vectors x and y.
- The ceiling value of $x \in \mathbb{R}$, denoted by $\lceil x \rceil$, is the smallest integer greater than x.

For a given set $S \in \mathbb{R}^n$:

- $\bar{s} = \max_{s \in S} s$ denotes the elementwise maximum where each element is computed as $\bar{s}_i = \max_{s \in S} s_i$. The elementwise minimum, $\underline{s}$ is defined similarly.
- For a matrix $A \in \mathbb{R}^{n \times m}$ we define the set $AS = \{z \in \mathbb{R}^n : z = Ax, \forall x \in S\}$.
- $\bar{S}$ denotes the complement of the set S which refers to elements not in (that is, elements outside of) the set S.

- cl(S) denotes the closure of set S which is defined as the union of S and its boundary.
- card(S) denotes the cardinality of a set S which is a measure of the number of elements of the set.
- $\text{Cone}(x, S) = \{x + \alpha(x - s), \forall s \in S, \forall \alpha \geq 0\}$ denotes the pointed cone with extreme point x and tangent to set S.

For given sets $X, Y \in \mathbb{R}^n$:

- $\text{Conv}(X, Y) = \{\alpha x + (1 - \alpha)y, \forall x \in X, \forall y \in Y, 0 \leq \alpha \leq 1\}$ is the convex hull of the sets X and Y.
- $X \cap Y$ denotes the set intersection between X and Y.
- $X \subset (\subseteq)Y$ denotes that X is a (strict) subset of Y.
- $X \oplus Y = x + y : x \in X,\ y \in Y$ defines the Minkowski addition of sets X and Y.
- $X \ominus Y = x \in X : x \oplus Y \subseteq X$ defines the Pontryagin difference of sets X and Y.
- The collection of all possible N combinations of binary variables is denoted by: $\{0, 1\}^N = \{(b_1 \ldots b_N) : b_i\{0, 1\}, i = 1 \ldots N\}$.
  The same definition holds for sign tuples $\{-, +\}^N$.
- For a binary variable f with values in $\{0, 1\}$, notation $\bar{f}$ denotes $\bar{f} = 1 - f$. The same holds for $f \in \{-, +\}$ where $\bar{f} = '-'$ if $f = '+'$ and $\bar{f} = '+'$ if $f = '-'$.
- $\text{lp}(n, d)$ and $\text{qp}(n, d)$ represent the complexity of solving a linear program, quadratic program respectively, with n constraints and d variables.

Chapter 1
Introduction

This book focuses on a class of control problems that can be translated to an optimization-based decision over a feasible region which is neither convex nor compact. This is the case in a wide range of problems which make use of non-convex constraints in conjunction with dynamical systems. More precisely, this issue arises naturally in many control engineering problems which put together *contradictory objectives*: persisting exciting control [1] (with both stabilization and regulation objectives and sufficiently large excitation in view of parameter identification), robust fault detection [2] (tracking error minimization against fault detection conditions) or the control of multi-agent systems [3] (path tracking against collision and obstacle avoidance conditions). In all these situations (and many other besides them), the presence of adversary constraints will result in non-convex feasible regions in which a constrained optimal solution has to be searched.

It is important to emphasize that these constraints are not just an artifact, but rather an intrinsic property which cannot be avoided. For example for obstacle avoidance, the constraints are handled explicitly in the mixed-integer formulations [4, 5], and implicitly in the Potential Field formulations [6], but regardless of the approach, they are always present. Another straightforward example of non-convex admissible domains is the one proposed in [7] where a "pit-jumping dilemma" is presented: a biker has to manage the speed to safely jump over the pit or avoid falling into it. This imposes some speed constraint, namely, either a minimum speed has to be obtained at the pit's edge in order to jump safely or a maximum speed has to be respected in order to stop in time. In other words, the bike's speed has to be outside a pre-defined interval. Formally, this translates into a non-convex constraint and therefore, the feasible region of decisions is non-convex (the complement of a forbidden speed interval). Another classical example which is illustrated in Fig. 1.1 deals with a boat which needs to cross from one bank to another given a certain river flow while avoiding small islands (see other examples in the viability framework [8]). As a consequence, in this case, non-convex position constraints need to be taken into

© The Author(s) 2016

I. Prodan et al., *Mixed-Integer Representations in Control Design*,
SpringerBriefs in Control, Automation and Robotics,
DOI 10.1007/978-3-319-26995-5_1

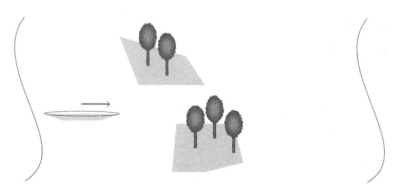

Fig. 1.1 Illustration of a boat crossing from one bank to another

account (such that the boat does not collide and safely avoids the islands). Beyond their particularities, similar questions arise in both examples: *how to determine such forbidden regions and safely avoid them? how to efficiently describe the feasible region from which the boat, bike, or in general a dynamical system is capable to steer or control trajectories away from a forbidden region?*

One of the popular approaches in tackling the enumeration and selection among alternatives in these problems is represented by *mixed-integer optimization* for which part or all of the arguments are required to be integers. For many years, MIP (Mixed-Integer Programing) problems have been actively bridging the gap between academia, industrial as well as business fields of life. The survey [9] lists several contributions in mixed-integer programming going back to the early 1820s. However, due to the inherent difficulty of these problem formulations, they have been used rather sporadically since. From a historical perspective, mathematicians have started first to analyze problems with integer variables. Important contributions were brought by Fourier [10] in 1826 who developed an algorithm for solving such problems by eliminating variables one at a time. Next, Farkas [11] in 1894 and Minkowski [12] in 1896 defined the polyhedral sets in a rigorous manner and their use in solving mixed-integer linear programs. Advancing throughout history, Dantzig, Gomory, Fulkerson, Hoffman and Kruskal were the pioneers in developing of combinatorial algorithm formulations like *simplex method* used to solve large scale applications [13] or other specific mixed-integer problems like the *traveling salesman problem* and *task assignment problems* in general [14–17]. Although mixed-integer programming is known to be NP-hard[1] in general, questions about polynomial time algorithms were first addressed by Edmonds [19] in 1965 who provided some efficient algorithms for Gaussian elimination method [20], and later by Garey and Johnson in 1979 [21, 22] and Karmarkar in 1984 [23], who provided various algorithms, mostly based

---

[1]NP-hard (Non-deterministic Polynomial-time hard), in computational complexity theory, is a class of problems that are, informally, "at least as hard as the hardest problems in NP". More precisely, a decision problem H is NP-hard when for any problem L in NP, there is a polynomial-time reduction from L to H [18].

on polyhedral approaches. Research on these issues are currently still very active. Furthermore, mixed-integer formulations, especially mixed-integer linear programming, due to their rigorousness, flexibility and extensive modeling capabilities, have become some of the most widely explored approaches for chemical process scheduling problems [24, 25].

With the increasing availability of computing power, interest in optimization problems which can be formulated through the use of mixed-integer techniques has been strongly increasing in the early 2000. As already mentioned, this method can be very useful in several fields of applications, due to its ability to include non-convex constraints and discrete decisions in the optimization problem. For example, in control engineering applications, collision avoidance plays an important role in the context of managing multiple agents (see, for instance, [3, 26], where different frameworks for the cooperative control of multiple agents are described). In the same time, it is known to be a difficult problem, since in general, not all the constraints are convex, [27]. In particular, the evolution of a dynamical system in an environment presenting obstacles can be modeled in terms of a non-convex feasible region. More precisely, it is possible to set up an optimization problem such that the agent state trajectory avoids a convex region, in fact, representing an obstacle (static constraints) or another agent (dynamic constraints-leading to a parameterization of the set of constraints with respect to the current state). This idea finds a natural formulation in the so-called predictive control formulation which is a receding horizon optimization based control design [28–30].

Consequently, it became obvious that solving optimization problems over non-convex regions is not a new issue in the control theory literature and, it is well known that mixed-integer formulations provide one of the best ways of dealing with this type of problems [9], at least from the point of view of the feasible domain representation. The advantage is that the problem is reformulated into a mixed continuous plus binary form which, despite the worst case combinatorial complexity, turns out to be manageable for reasonable dimensions and furthermore permits the use of efficient algorithms.

Remarkably, contributions in control area have focused more on optimizing agent trajectories [31–33]. For example, multi-vehicle target assignment and intercept problems were studied intensively by Earl and D'Andrea [27], Beard et al. [34]. Next, mixed-integer techniques were also useful for coordinating the efficient interaction of multiple agents in scenarios with many sequential tasks and tight timing constraints (see, [35, 36]). Bemporad et al. [37, 38] used a combination of mixed-integer formulation and Model Predictive Control (MPC) to stabilize general hybrid systems around equilibrium points. Other recent work [39] introduced mixed-integer programming in a predictive control framework to plan short trajectories around nearby obstacles. The mixed-integer formulation has also proven to be useful for the convergence of multiple agents towards a tight formation as well as in various distributed predictive control problems [4, 40, 41]. In a different control framework, a design application was reported in [42], where a feasible reference signal allowing set membership testing for fault detection was computed over a non-convex region, leading to a mixed-integer formulation. Also, it is worth mentioning the use of mixed-integer constructions in *bilevel problems*. Rewriting the bilevel problem by replacing the lower-level optimization problem with its dual form (through the use of Karas-Kuhn-Tucker (KKT) conditions) leads to non-convex complementarity constraints which can be described through the mixed-integer formalism [43, 44].

However, when discussing aspects like modeling capabilities it is important to note from the beginning that in the mixed-integer formulation, the computational complexity is highly dependent on the size of the binary part which limits its usefulness to relatively small-size problems (even for linear and quadratic cost function). Consequently, these methods may not be "fast" enough for real-time control of systems with large problem formulations. There has been a number of attempts in the literature to reduce the computational requirements of MIP formulations in order to make them attractive for real-time applications. In [45], an iterative method for including the obstacles in the best path generation is provided. Other references, like [46], consider a predefined path constrained by a sequence of convex sets. In all of these works, the binary variables reduction is not tackled at the MIP level, but, instead, the original decision problems are reformulated in a simplified MIP form.

The negative influence of the large number of binary variables in the problem formulation highlights the importance of reducing them. There are several contributions focusing on reducing the number of binary variables used in the problem formulation. For example, Vielma and Nemhauser in [47] (and the references therein) discussed a logarithmic formulation. This, and other techniques help to reduce the computational burden but, ultimately, the complexity of the problem is directly related to the difficulty of representing the feasible region [48, 49].

As it is often the case, a careful analysis of the problem "at-hand" can improve the outcome of the result both quantitatively (reduced computation time, efficient formulations etc.) and qualitatively (observing that the underlying structure may lead to conceptual breakthroughs and opens new research perspectives). While various tools for describing non-convex and non-connected regions exist, we maintain that they are usually unnecessarily complex and that by acknowledging the combinatorial nature of the problem, more efficient formulations can be reached.

Roughly speaking, the main idea used throughout the book can be resumed as follows: using notions like hyperplane arrangements, cell merging and Boolean algebra, the number of binary variable appearing in a mixed-integer optimization problem is reduced via a more efficient coding. Furthermore, the use of the combinatorial notion of hyperplane arrangements will be particularly helpful for the problems to be considered (see for example the work of Zaslavsky, Orlik and others [50–54] for a detailed mathematical description and various applications of hyperplane arrangements notions). Such a construction defines the partitioning of the space in a finite collection of hyperplanes and can be used to describe efficiently non-convex regions. It is worth mentioning that, in general, throughout the book, the class of convex sets used is represented by polyhedral sets since, in our opinion, they represent a good balance between representation complexity and flexibility with respect to various algebraic operations (addition, intersection and the like). Indeed, it is well known that any convex and compact set can be arbitrarily closely approximated by a polyhedron [55].

With such a construction in place, we can then apply mixed-integer programming to put the overall problem into an acceptable formulation which can be further solved by specialized solvers. Note that some of the associated mixed-integer constructions are uniquely fitted to the hyperplane arrangement construction and make full use of its combinatorial nature. Briefly, some of the noteworthy aspects of the approach which resumes also the main contributions of this book are the following:

- a convex representation in the extended space of the finite state vector of the dynamical system plus binary variables using the associated hyperplane arrangement;
- reduced complexity of the mixed-integer problem formulation upon Boolean algebra techniques;
- a notable property of optimal association between regions and their binary representation leading to the minimization of the number of constraints.

These advances can be showcased in a variety of control problems. Without covering exhaustively the possible applications, we have chosen here to exemplify the issues afflicting multi-agent constructions with obstacles (i.e., collision and avoidance constraints, the coverage problem) and an active FTC (Fault Tolerant Control) scheme based on invariant set separation. More precisely, for the first class of applications, the agent, which may have an associated polyhedral safety region, is required to maneuver successfully in some hostile environment. Without loss of generality, the obstacles are designed as convex polyhedral regions. For the second class of applications, the main purpose of the FTC is to automatically attenuate/cancel the negative effects of a component fault. This requires persistent excitation in the dynamical system such that FDI (Fault Detection and Isolation) is attainable. In other words, the feasible domain is the complement of a neighborhood centered around the origin.

Since multiple constructions are possible starting from the same initial problem, the complexity of the resulting constructions under various parameters (space dimension, number of binary variables, etc.) will be also discussed.

It is important to underline that we do not address the solver issue. Although such an aspect is extremely important, it is out of the scope of the present monograph. In this context, it is important to further mention that the state of the art of MIP solvers are in many cases remarkably effective, and have improved radically in the last years. These solvers typically use branch-and-cut algorithms [56], and cutting planes algorithms [57] to obtain improved linear programming bounds and branching to carry out implicit enumeration of the solutions. It turns out that Gomory mixed-integer cuts are one of the most crucial ingredients for the success of current mixed-integer optimization solvers [58]. A detailed survey on the mixed-integer programming computation is provide by Lodi in [59] (see also Chap. 16 of the monograph [9]). The availability of commercial as well as non-commercial MIP solvers has created over the past decades and continues to create nowadays a very fruitful competition stimulating the entire MIP community to go "larger" and "faster".

Among others, a well know commercial solver, originally founded by Robert Bixby [60], is CPLEX which started in the 90s with version CPLEX 1.2 and arrived now, in 2015, at version CPLEX 12.6.2. Bixby et al. [61] report that in 2004 an LP (Linear Programming) problem was solved, by CPLEX 8, a million times faster than it was by CPLEX 1 in 1990, three orders of magnitudes due to hardware and to software improvements, respectively. This gives a clear (however limited) indication of how much LP technology has been and still is important for MIP development. Next, Gurobi Optimizer [62], named after its founders, Gu Zonghao, Rothberg Edward and Bixby Robert represents a more recent commercial optimization solver for LP, QP (Quadratic Programming), QCP (Quadratically Constrained Programming), MILP (Mixed-Integer Linear Programming), MIQP (Mixed-Integer Quadratic Programming), and MIQCP (Mixed-Integer Quadratically Constrained Programming). The solver SCIP (Solving Constraint Integer Programs) developed by Tobias Achterberg [63, 64] is currently an example of non-commercial MIP solver which can handle mixed-integer programming and MINLP (Mixed-Integer Nonlinear Programming) problems [59, 65]. Finally, we mentioned that there are still very difficult classes of MIPs on which the current solvers are not effective as the complex scheduling problems in operations research [66].

To summarize, the goal in this book is to show ways in which, given an initial formulation of control or decision problem, its structure can be used to obtain improved optimization-based formulations and expect for more effective algorithms that take the structure into account.

The reminder of the book is organized in three chapters covering prerequisites, basic notions, mixed-integer representations as well as various examples. Chapter 2 provides the necessary preliminaries and prerequisites (polyhedral sets, hyperplane arrangements and the like) which are then used to describe feasible regions under various methodologies. In particular, the efficient description of the region and minimization of run-time computations are emphasized. Chapter 3 considers the previous geometrical constructions and puts them under the mixed-integer formalism. In this context, we provide several different constructions and discuss their various particularities which makes them more or less adapted to the control problem at hand. Next, Chap. 4 presents various applications of the mixed-integer constructions for representative control problems. First, a multi-agent control problem and various issues which may affect it (collision and avoidance constraints, coverage constraints and the like) is considered. Second, an active FTC scheme which reacts to a detected fault and reconfigures the control actions is described in detail. All along these main chapters, proof-of-concept illustrations are provided and, in some cases, numerical issues are presented and discussed. The conclusions are drawn in Chap. 5.

Finally, we are convinced that the approaches and tools presented in this book will provide a valuable support to a better understanding of the analysis and control of dynamical systems with conflicting objectives and an effective usage of the associated mixed-integer formulations.

References

1. Marafioti, G., Stoican, F., Bitmead, R.R., Hovd, M.: Persistently exciting model predictive control for siso systems. In: 4th IFAC Nonlinear Model Predictive Control Conference International Federation of Automatic Control, pp. 448–453. Noordwijkerhout, Netherlands (2012)
2. Stoican, F., Olaru, S., Seron, M., De Doná, J.: Reference governor design for tracking problems with fault detection guarantees. J. Process Control **22**(5), 829–836 (2012)
3. Grundel, D., Murphey, R., Pardalos, P.: Cooperative Systems, Control and Optimization, vol. 588. Springer (2007)
4. Prodan, I., Olaru, S., Stoica, C., Niculescu, S.I.: On the tight formation for multi-agent dynamical systems. In: KES—Agents and Multi-agent Systems—Technologies and Applications, pp. 554–565. Springer (2012)
5. Prodan, I., Bitsoris, G., Olaru, S., Stoica, C., Niculescu, S.: On the limit behavior for multi-agent dynamical systems. In: The IFAC Workshop on Navigation, Guidance and Control of Underwater Vehicles, pp. 106–111. Porto, Portugal (2012)
6. Prodan, I., Olaru, S., Stoica, S., Niculescu, S.I.: Predictive control for trajectory tracking and decentralized navigation of multi-agent formations. Int. J. Appl. Math. Comput. Sci. **23**(1), 91–102 (2013)
7. Blanchini, F., Miani, S.: Set-Theoretic Methods in Control, 2nd edn. Springer (2014)
8. Aubin, J.: Viability Theory. Birkhauser, Boston (1991)
9. Jünger, M., Junger, M., Liebling, T., Naddef, D., Nemhauser, G., Pulleyblank, W.: 50 Years of Integer Programming 1958-2008: From the Early Years to the State-of-the-Art. Springer (2009)
10. Fourier, J.: Solution d'une question particulière du calcul des inégalités. Nouveau Bulletin des Sciences par la Société Philomatique de Paris, pp. 317–319 (1826)
11. Farkas, G.: On the applications of the mechanical principle of fourier. Mathematikai és Természettudományi Értesoto **12**, 457–472 (1894)
12. Minkowski, H.: Geometrie der Zahlen (Erste Lieferung). Teubner, Leipzig (1896)
13. Dantzig, G.: Maximization of a linear function of variables subject to linear inequalities. In: Koopmans, T.C. (ed.) Activity Analysis of Production and Allocation, pp. 339–347. Wiley, New York (1951)
14. Dantzig, G., Fulkerson, R., Johnson, S.: Solution of a large-scale traveling-salesman problem. J. Oper. Res. Soc. Am. **2**(4), 393–410 (1954)
15. Kuhn, H.W.: The hungarian method for the assignment problem. Naval Res. Logist. Q. **2**(1–2), 83–97 (1955)
16. Hoffman, A.J., Kruskal, J.B.: Integral boundary points of convex polyhedra. In: 50 Years of Integer Programming 1958–2008, pp. 49–76. Springer (2010)
17. Gomory, R.E., et al.: Outline of an algorithm for integer solutions to linear programs. Bull. Am. Math. Soc. **64**(5), 275–278 (1958)
18. Van Leeuwen, J., Leeuwen, J.: Handbook of Theoretical Computer Science: Algorithms and Complexity, vol. 1. Elsevier (1990)
19. Edmonds, J.: Paths, trees, and flowers. Can. J. Math. **17**, 449–467 (1965)
20. Edmonds, J.: Systems of distinct representatives and linear algebra. J. Res. Nat. Bur. Stan. B **71**, 241–245 (1967)
21. Garey, M., Johnson, D.: Computers and intractability. A guide to the theory of NP-completeness. A Series of Books in the Mathematical Sciences. WH Freeman and Company, San Francisco, Calif (1979)
22. Khachiyan, L.: A polynomial algorithm in linear programming. Sov. Math. Dokl. **20**, 191–194 (1979)
23. Karmarkar, N.: A new polynomial-time algorithm for linear programming. Combinatorica **4**, 373–395 (1984)
24. Pritsker, A.A.B., Waiters, L.J., Wolfe, P.M.: Multiproject scheduling with limited resources: a zero-one programming approach. Manag. Sci. **16**(1), 93–108 (1969)

25. Sahinidis, N., Grossmann, I.E.: Reformulation of multiperiod milp models for planning and scheduling of chemical processes. Comput. Chem. Eng. **15**(4), 255–272 (1991)
26. Grundel, D., Pardalos, P.: Theory and Algorithms for Cooperative Systems, vol. 4. World Scientific Publishing Co Inc. (2004)
27. Earl, M., D'Andrea, R.: Modeling and control of a multi-agent system using mixed integer linear programming. In: Proceedings of the 40th IEEE Conference on Decision and Control, vol. 1, pp. 107–111. Orlando, Florida, USA (2001)
28. Propoi, A.: Use of linear programming methods for synthesizing sampled-data automatic systems. Autom. Remote Control **24**(7), 837–844 (1963)
29. Richalet, J., Rault, A., Testud, J., Papon, J.: Model predictive heuristic control: applications to industrial processes. Automatica **14**(5), 413–428 (1978)
30. Mayne, D.Q., Schroeder, W.: Robust time-optimal control of constrained linear systems. Automatica **33**, 2103–2118 (1997)
31. Schouwenaars, T., De Moor, B., Feron, E., How, J.: Mixed integer programming for multi-vehicle path planning. In: Proceedings of the 2nd IEEE European Control Conference, pp. 2603–2608. Citeseer, Porto, Portugal (2001)
32. Richards, A., How, J.: Model predictive control of vehicle maneuvers with guaranteed completion time and robust feasibility. In: Proceedings of the 24th American Control Conference, vol. 5, pp. 4034–4040. Portland, Oregon, USA (2005)
33. Ousingsawat, J., Campbell, M.: Establishing trajectories for multi-vehicle reconnaissance. In: Proceedings of the 22nd AIAA Guidance, Navigation, and Control Conference, pp. 2188–2199. Providence, Rhode Island, USA (2004)
34. Beard, R., McLain, T., Goodrich, M., Anderson, E.: Coordinated target assignment and intercept for unmanned air vehicles. Proc. IEEE Trans. Robot. Autom. **18**(6), 911–922 (2002)
35. Richards, A., Bellingham, J., Tillerson, M., How, J.: Coordination and control of multiple UAVs. In: AIAA Guidance, Navigation, and Control Conference, Monterey, CA (2002)
36. Schumacher, C.: UAV task assignment with timing constraints. Tech. rep., Air Force research lab Wright-Patterson AFB of air vehicles directorate (2003)
37. Bemporad, A., Morari, M.: Control of systems integrating logic, dynamics, and constraints. Automatica **35**, 407–428 (1999)
38. Bemporad, A., Borrelli, F., Morari, M.: Optimal controllers for hybrid systems: stability and piecewise linear explicit form. In: Proceedings of the 39th IEEE Conference on Decision and Control, vol. 2, pp. 1810–1815. IEEE (2000)
39. Bellingham, J., Richards, A., How, J.: Receding horizon control of autonomous aerial vehicles. In: IEEE (IEEE (ed.): Proceedings of the 21th American Control Conference, pp. 138–143. Anchorage, Alaska, USA (2002)
40. Prodan, I., Olaru, S., Stoica Maniu, C., Niculescu, S.I.: Predictive control for tight group formation of multi-agent systems. In: Proceedings of the 18th IFAC World Congress. Milano, Italy (2011), pp. 138–143
41. Prodan, I., Stoican, F., Olaru, S., Stoica, C., Niculescu, S.I.: Mixed-integer programming techniques in distributed MPC problems. In: Distributed MPC Made Easy, vol. 69, pp. 273–288. Springer (2013)
42. Stoican, F., Olaru, S., Seron, M., De Doná, J.: Reference governor for tracking with fault detection capabilities. In: Proceedings of the 2010 Conference on Control and Fault Tolerant Systems, pp. 546–551. Nice, France (2010)
43. Hovd, M.: Multi-level Programming for Designing Penalty Functions for MPC Controllers. In: Proceedings of the 18th IFAC World Congress, pp. 6098–6103. Milano, Italy (2011)
44. Hovd, M., Stoican, F.: On the design of exact penalty functions for MPC using mixed integer programming. Comput. Chem. Eng. **70**, 104–113 (2013)
45. Earl, M.G., D'Andrea, R.: Iterative MILP methods for vehicle-control problems. IEEE Trans. Robot. **21**(6), 1158–1167 (2005)
46. Vitus, M.P., Pradeep, V., Hoffmann, G., Waslander, S.L., Tomlin, C.J.: Tunnel-milp: path planning with sequential convex polytopes. In: AIAA Guidance, Navigation, and Control Conference, pp. 1–13. Honolulu, Hawaii, USA (2008)

47. Vielma, J., Nemhauser, G.: Modeling disjunctive constraints with a logarithmic number of binary variables and constraints. Math. Prog. **128**(1), 49–72 (2011)
48. Stoican, F., Prodan, I., Olaru, S.: On the hyperplanes arrangements in mixed-integer techniques. In: Proceedings of the 30th American Control Conference, pp. 1898–1903. San Francisco, California, USA (2011)
49. Stoican, F., Prodan, I., Olaru, S.: Enhancements on the hyperplane arrangements in mixed integer techniques. In: Proceedings of the 50th IEEE Conference on Decision and Control and European Control Conference, pp. 3986–3991. Orlando, Florida, USA (2011)
50. Zaslavsky, T.: Facing up to arrangements: face-count formulas for partitions of space by hyperplanes. American Mathematical Society (1975)
51. Edelsbrunner, H., Seidel, R., Sharir, M.: On the zone theorem for hyperplane arrangements. In: New Results and New Trends in Computer Science, pp. 108–123 (1991)
52. Orlik, P., Terao, H.: Arrangements of Hyperplanes, vol. 300. Springer (1992)
53. Geyer, T., Torrisi, F., Morari, M.: Optimal complexity reduction of piecewise affine models based on hyperplane arrangements. In: Proceedings of the 23th American Control Conference, vol. 2, pp. 1190–1195. Boston, Massachusetts, USA (2004)
54. Orlik, P.: Hyperplane arrangements. In: Floudas, C., Pardalos, P. (eds.) Encyclopedia of Optimization, pp. 1545–1547. Springer, US (2009)
55. Cairns, S.S.: Polyhedral approximations to regular loci. Ann. Math. 409–415 (1936)
56. Padberg, M., Rinaldi, G.: A branch and cut algorithm for the resolution of large-scale symmetric traveling salesmen problems. SIAM Rev. **33**, 60–100 (1991)
57. Land, A., Doig, A.: An automatic method of solving discrete programming problems. Econometrica **28**, 497–520 (1960)
58. Gomory, R.: Outline of an algorithm for integer solutions to linear programs. Bull. Am. Math. Soc. **64**, 275–278 (1958)
59. Lodi, A.: Mixed integer programming computation. In: 50 Years of Integer Programming 1958–2008, pp. 619–645. Springer (2010)
60. Bixby, R.: Solving real-world linear programs: a decade and more of progress. Oper. Res. **50**, 315 (2002)
61. Bixby, R.: Mixed-integer programming: a progress report, the sharpest cut: the impact of manfred padberg and his work. In: Grotschel, M. (ed.) MPS-SIAM Series on Optimization pp. 309–325 (2004)
62. Optimization, G.: Gurobi Optimizer Reference Manual. http://www.gurobi.com (2012)
63. Achterberg, T.: SCIP-a framework to integrate constraint and mixed integer programming. Konrad-Zuse-Zentrum für Informationstechnik (2004)
64. Achterberg, T.: Scip: solving constraint integer programs. Math. Prog. Comput. **1**(1), 1–41 (2009)
65. D'Ambrosio, C., Lodi, A.: Mixed integer nonlinear programming tools: a practical overview. 4OR **9**(4), 329–349 (2011)
66. Borghetti, A., D'Ambrosio, C., Lodi, A., Martello, S.: An milp approach for short-term hydro scheduling and unit commitment with head-dependent reservoir. IEEE Trans. Power Syst. **23**(3), 1115–1124 (2008)

Chapter 2
Non-convex Region Description by Hyperplane Arrangements

This chapter presents some prerequisites and basic notions which will be instrumental in the rest of the manuscript. Namely, details about hyperplane arrangements (the scaffolding over which the feasible region characterization is defined) and various standard set theoretic elements (families of sets like polyhedra, zonotopes as well as basic set operations) are presented.

Without being exhaustive, for describing this "set of tools" the presentation follows the ideas and notations in [1–3]. Furthermore, these notions can be readily found and reviewed by the interested readers in complementary materials referenced in this chapter.

2.1 Hyperplane Arrangements

Let us consider a finite collection of hyperplanes $\mathbb{H} = \{\mathcal{H}_i\}_{i \in \mathbb{I}}$ from $\mathbb{R}^n$:

$$\mathcal{H}_i = \left\{ x \in \mathbb{R}^n : h_i x = k_i \right\}, \ i \in \mathbb{I}, \tag{2.1}$$

with $\mathbb{I} \triangleq \{1 \ldots N\}$ and $(h_i, k_i) \in \mathbb{R}^{1 \times n} \times \mathbb{R}$.

Each of these hyperplanes partitions the space into two disjoint[1] regions (which halve the space and hence are called "half-spaces"):

$$\mathcal{R}_i^+ = \left\{ x \in \mathbb{R}^n : \quad h_i x \leq \quad k_i \right\}, \tag{2.2a}$$
$$\mathcal{R}_i^- = \left\{ x \in \mathbb{R}^n : -h_i x \leq -k_i \right\}. \tag{2.2b}$$

---

[1] The relative interiors of these regions do not intersect, but their closures have as common boundary the affine subspace $\mathcal{H}_i$ in (2.1).

© The Author(s) 2016
I. Prodan et al., *Mixed-Integer Representations in Control Design*,
SpringerBriefs in Control, Automation and Robotics,
DOI 10.1007/978-3-319-26995-5_2

Considering the above basic concepts the combinatorial notion of hyperplane arrangement is defined in the following.

Definition 2.1 (*Hyperplane arrangements*—[4]) The collection of hyperplanes $\mathbb{H}$ will partition the space into a union of disjoint cells $\mathscr{A}(\sigma)$ characterized by a sign tuple $\sigma \in \{-, +\}^N$ defined as follows:

$$\mathscr{A}(\sigma) = \bigcap_{i \in \mathbb{I}} \mathscr{R}_i^{\sigma(i)}. \tag{2.3}$$

The collection of all feasible sign tuples describes a hyperplane arrangement of feasible cells covering the entire space:

$$\mathscr{A}(\mathbb{H}) = \bigcup_{\sigma \in \Sigma} \mathscr{A}(\sigma), \tag{2.4}$$

where

$$\Sigma = \{\sigma \in \{-, +\}^N : \mathscr{A}(\sigma) \neq \emptyset\}, \tag{2.5}$$

denotes the collection of sign tuples resulting into non-empty intersections of regions (2.2a)–(2.2b).   ◆

Remark 2.1 Arrangement (2.4) is said to be in *general position* if for any 'i' a perturbation $\epsilon_i > 0$ (e.g., $\mathscr{H}_i \to \mathscr{H}_i' = \{x : h_i x = k_i + \epsilon_i\}$) will not change the distribution and number of regions.   ◆

Remark 2.2 Any sub-arrangement $\mathscr{A}(\mathbb{H}')$ of $\mathscr{A}(\mathbb{H})$ (i.e., $\mathbb{H}' \subseteq \mathbb{H}$) is said to be *central* if $\bigcap_{\mathscr{H} \in \mathbb{H}'} \mathscr{H} \neq \emptyset$.   ◆

Illustrative example for hyperplane arrangements in $\mathbb{R}^2$

Here and thereafter, we make use of the numerical data provided in Appendix A such that the present and the forthcoming illustrative examples are based on reproducible constructions.

Figure 2.1a depicts a hyperplane as described by (2.1), $\mathscr{H} = \{x : [0.07\ 0.99] x = 4.98\}$ and its two half-spaces, $\mathscr{R}^- = \{x : [0.07\ 0.99] x \leq 4.98\}$ and $\mathscr{R}^+ = \{x : [0.07\ 0.99] x \geq 4.98\}$ as in (2.2a)–(2.2b).

Figure 2.1b illustrates 4 hyperplanes with the numerical data provided in Appendix A.1. These generate a hyperplane arrangement as in (2.3) with 10 cells, each defined by a unique sign tuple (the feasible sign combinations are: $(-+--)$, $(----)$, $(+---)$, $(+-+-)$, $(+--+)$, $(+-++)$, $(++++)$, $(-+++)$, $(-++-)$, $(+++-)$). Also, the figure delineates the cell $\mathscr{A}(\sigma) = \bigcup_{\sigma(i)} \mathscr{R}_i^{\sigma(i)} = \mathscr{R}_1^+ \cap \mathscr{R}_2^+ \cap \mathscr{R}_3^+ \cap \mathscr{R}_4^-$, corresponding to the sign tuple $\sigma = (+++-)$. Note that all the other sign tuples (the remaining $2^4 - 10 = 6$ sign combinations) are infeasible. For example, $\sigma = (-+-+)$ which corresponds to $\mathscr{A}(\sigma) = \bigcup_{\sigma(i)} \mathscr{R}_i^{\sigma(i)} = \mathscr{R}_1^- \cap \mathscr{R}_2^+ \cap \mathscr{R}_3^- \cap \mathscr{R}_4^+$ is an empty set (due to the fact that $(\mathscr{R}_1^- \cap \mathscr{R}_2^+ \cap \mathscr{R}_3^-) \cap \mathscr{R}_4^+ = \emptyset$).

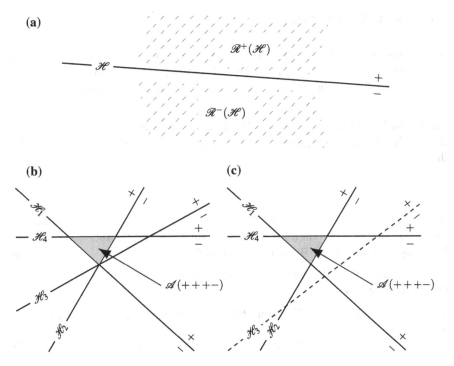

Fig. 2.1 Hyperplane arrangements. **a** One hyperplane and its associated half-spaces. **b** Hyperplane arrangement. **c** Perturbed hyperplane arrangement

Furthermore, note that the hyperplane arrangement depicted in Fig. 2.1b is not in a general position since three of the hyperplanes intersect $\{\mathcal{H}_1, \mathcal{H}_2, \mathcal{H}_3\}$. These three hyperplanes also define a central sub-arrangement but no other 3 hyperplanes define one (e.g., $\{\mathcal{H}_1, \mathcal{H}_2, \mathcal{H}_4\}$), as explained in Remark 2.2. Note that any combination of tho hyperplanes is a central sub-arrangement ($\{\mathcal{H}_1, \mathcal{H}_2\}$, $\{\mathcal{H}_1, \mathcal{H}_3\}$, $\{\mathcal{H}_1, \mathcal{H}_4\}$, $\{\mathcal{H}_2, \mathcal{H}_4\}$, $\{\mathcal{H}_3, \mathcal{H}_4\}$).

Figure 2.1c delineates the arrangement perturbed by translating the 3rd hyperplane ($\mathcal{H}_3 \rightarrow \{[0.6 \ -0.8]\, x = 1\}$). The corresponding numerical data is to be found in Appendix A.1. We observe that no three hyperplanes intersect in a common point and that the number of cells has increased to 11. This was expected since in $\mathbb{R}^2$ three randomly taken hyperplanes should not intersect in a common point (see Remark 2.1).

2.1.1 Region Counting

Each of the regions (2.4) can be either bounded or unbounded. Hence, denote[2] the total number of regions as $r(\mathscr{A})$ and the number of bounded regions as $b(\mathscr{A})$.

To proceed further, the auxiliary notion of a *characteristic polynomial* is required.

Definition 2.2 (*Whitney's formula* [5]) Let there be an arrangement $\mathscr{A}$ in $\mathbb{R}^n$ and $\mathscr{A}'$ the shorthand notation for any of its sub-arrangements (as defined in Remark 2.2), then:

$$\Xi(\mathscr{A}, t) = \sum_{\mathscr{A}' \subseteq \mathscr{A}} (-1)^{\mathrm{card}(\mathscr{A}')} t^{n-\mathrm{rank}(\mathscr{A}')}, \qquad (2.6)$$

where $\mathscr{A}'$ are central sub-arrangements of $\mathscr{A}$, $\mathrm{card}(\mathscr{A}')$ denotes the number of hyperplanes of $\mathscr{A}'$ and $\mathrm{rank}(\mathscr{A}')$ denotes the dimension of the sub-space spanned by the hyperplanes of $\mathscr{A}'$. By definition, $\mathrm{rank}(\emptyset) = 0$. ◆

The number of regions (total and bounded) can now be given as a function of the characteristic polynomial.

Theorem 2.1 (*Zaslavsky's Theorem* [6]) *For a given arrangement $\mathscr{A}$, the number of regions is*

$$r(\mathscr{A}) = |\Xi(\mathscr{A}, -1)|, \qquad b(\mathscr{A}) = |\Xi(\mathscr{A}, 1)|. \qquad (2.7)$$

□

Remark 2.3 The above notions provide powerful tools for counting the regions (which is of interest as it reflects on the number of binary variables required in the mixed-integer constructions of the next chapter). Note that all the computations involve simple operations and avoid explicitly computing the regions (hence one can proceed with combinatorial notions and avoid a costly geometrical analysis). Furthermore, whenever the arrangement is in general position (see Remark 2.1) these numbers are known. For example,

$$r(\mathscr{A}) = \sum_{i=0}^{n} \binom{N}{i}, \qquad (2.8)$$

as per Buck's formula [7]. ◆

Illustrative example for region counting

Let us revisit first the example in Fig. 2.1b where the hyperplanes are not in general position (as defined in Remark 2.1) and hence the bound given by Buck's formula is not reached. We proceed to enumerate all the central sub-arrangements of the arrangement: $\{\emptyset\}, \{\mathscr{H}_1\}, \{\mathscr{H}_2\}, \{\mathscr{H}_3\}, \{\mathscr{H}_4\}, \{\mathscr{H}_1, \mathscr{H}_2\}, \{\mathscr{H}_1, \mathscr{H}_3\}, \{\mathscr{H}_1, \mathscr{H}_4\}, \{\mathscr{H}_2, \mathscr{H}_3\},$

---

[2]Whenever the notation is not ambiguous, we use the shorthand $\mathscr{A}$ for the arrangement $\mathscr{A}(\mathbb{H})$.

$\{\mathcal{H}_2, \mathcal{H}_4\}$, $\{\mathcal{H}_3, \mathcal{H}_4\}$ and $\{\mathcal{H}_1, \mathcal{H}_2, \mathcal{H}_3\}$. To each of these central sub-arrangements corresponds a term in Whitney's formula. For example $\mathcal{A}' = \{\mathcal{H}_1, \mathcal{H}_4\}$ has card $(\mathcal{A}') = 2$ elements and rank$(\mathcal{A}') = 2$ (since the two composing hyperplanes span $\mathbb{R}^2$) thus leading to term $(-1)^2 t^{2-2} = 1$. Applying the same reasoning for the rest of the sub-arrangements enumerated before we obtain the characteristic polynomial (2.6):

$$\Xi(\mathcal{A}, t) = 1 \cdot (-1)^3 t^{2-2} + 6 \cdot (-1)^2 t^{2-2} + 4 \cdot (-1)^1 t^{2-1} + 1 \cdot (-1)^1 t^{2-0}$$
$$= 5 - 4t + t^2,$$

which permits to apply Zaslavsky's Theorem and obtain the total number of regions $r(\mathcal{A}) = |\Xi(\mathcal{A}, -1)| = 10$ and the number of bounded regions $b(\mathcal{A}) = |\Xi(\mathcal{A}, 1)| = 2$.

The same constructions can be applied to the arrangement from Fig. 2.1c. The only difference is that there are no longer three intersecting hyperplanes and hence the term $(-1)^3 t^{2-2}$ no longer appears in the characteristic polynomial:

$$\Xi(\mathcal{A}, t) = 6 \cdot (-1)^2 t^{2-2} + 4 \cdot (-1)^1 t^{2-1} + 1 \cdot (-1)^1 t^{2-0} = 6 - 4t + t^2,$$

In turn, by again applying the Zaslavsky's Theorem, the total number of regions $r(\mathcal{A}) = |\Xi(\mathcal{A}, -1)| = 11$ and the number of bounded regions $b(\mathcal{A}) = |\Xi(\mathcal{A}, 1)| = 3$ are obtained. For the total number of regions, we may as well apply Buck's formula, i.e., $r(\mathcal{A}) = \sum_{i=0}^{2} \binom{4}{i} = \binom{4}{0} + \binom{4}{1} + \binom{4}{2} = 11$.

The bounded and total number of regions are illustrated in Fig. 2.2 which uses the numerical data provided in Appendix A.1.

2.1.2 Parametrized Hyperplane Arrangements

In many practical control applications the hyperplane arrangement modifies from one instant to the next in relationship with a dynamical system evolution. Perhaps the most illustrative example is to consider a moving obstacle: the hyperplanes which characterize it will translate from one instant to the next. Re-computing the arrangement is not trivial and may be impractical for small discretization steps. The solution is to analyze the problem in a lifted space (variables and parameters), obtain the corresponding arrangement and to 'cut' it at runtime for the current parameter in order to go back to the original space.

Henceforth, let us consider a collection of parametrized hyperplanes $\mathbb{H}(p) = \{\mathcal{H}_i(p)\}$ from $\mathbb{R}^n$ where each of them has a linear dependence in the right-hand side with respect to a variable $p \in \mathbb{R}^{n_p}$:

$$\mathcal{H}_i(p) = \{x \in \mathbb{R}^n : h_i x = k_i - h_i^p p\}. \tag{2.9}$$

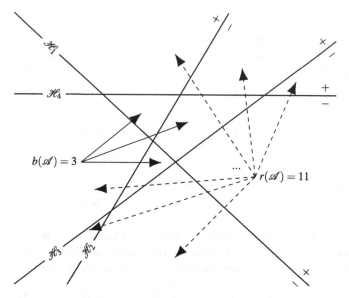

Fig. 2.2 Hyperplane arrangement region counting

A new hyperplane collection $\mathbb{H}' = \{\mathscr{H}_i'\}$ is considered:

$$\mathscr{H}_i' = \left\{ [h_i \; h_i^p] \cdot \begin{bmatrix} x \\ p \end{bmatrix} = k_i \right\}, \tag{2.10}$$

and the arrangement $\mathscr{A}(\mathbb{H}')$ is defined with feasible cells $\mathscr{A}(\sigma')$ where $\sigma' \in \Sigma'$.

To retrieve the cells composing the arrangement in $\mathbb{R}^n$ for a certain parameter p^* we simply intersect $\mathscr{A}(\sigma')$ with the subspace $\{p = p^*\}$:

$$\mathbb{H}(p^*) = \bigcup_{\sigma' \in \Sigma'} \left\{ x : A(\sigma') \cap \{p = p^*\} \right\}. \tag{2.11}$$

At this point we might stop if not for the fact that many of the cells $A(\sigma') \cap \{p = p^*\}$ will be empty for certain values of p. To reduce the runtime computations we can therefore consider an additional analysis step. That is, we compute the domain of existence of each cell $A(\sigma')$ in the parameter space (i.e., project on $\mathbb{R}^{n_p}$):

$$\mathrm{dom}(\mathscr{A}(\sigma')) = \left\{ p : \exists x \; s.t. \begin{bmatrix} x \\ p \end{bmatrix} \in \mathscr{A}(\sigma') \right\}. \tag{2.12}$$

This provides a partitioning of $\mathbb{R}^{n_p}$ into overlapping domains which can be further written (after suitable manipulations) as a union of disjoint domains D_k'. To each domain corresponds a list of sign tuples Σ_k' such that $\forall \sigma' \in \Sigma_k'$, $\exists p \in D_k'$ s.t. $\mathscr{A}(\sigma') \cap \{p = p^*\} \neq \emptyset$. Lastly, at runtime, we identify index 'k' such that

$p \in D_k$ and retrieve Σ'_k from which we generate the current arrangement, as in (2.11) but with Σ' reduced to Σ'_k.

Remark 2.4 These notions can be related to the parametrized polyhedra and associated validity domains (see Sect. 2.2 and [8]).   ◆

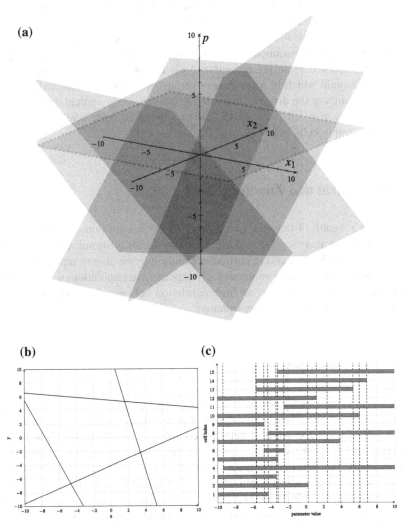

Fig. 2.3 Parametrized hyperplane arrangements. **a** Hyperplane arrangement lifted in $\mathbb{R}^3$. **b** Domain. **c** Region

Illustrative example for hyperplane parametrization

Let us consider the hyperplane arrangement given in Appendix A.2 with 4 hyperplanes in $\mathbb{R}^2$ parametrized after a scalar $p \in \mathbb{R}$. We apply (2.10) and obtain an arrangement with 4 hyperplanes in $\mathbb{R}^3$ which has (according to (2.8)) 15 cells.

In Fig. 2.3a we depict the resulting arrangement with the original space represented along dimensions x_1 and x_2 and the parameter space represented along dimension x_3. It can be seen that 'cutting' for a certain value of p will result in different arrangements in $\mathbb{R}^2$, see Fig. 2.3b which is obtained for $p = 1.5$.

Lastly, in Fig. 2.3c we illustrate the domains of existence for the cells of the lifted arrangement. Since the parameter is a scalar, these domains are intervals along the real axis (represented as solid segments). As it can be seen, each of the 15 cells has its own domain which in some cases overlaps with others. Considering all the overlaps we compute the disjoint intervals D'_k (separated by vertical dashed lines). For example, for $p = -1.5 \in [-2.45\ 2.45]$ we have 10 cells whose cutting will result in non-empty cells in $\mathbb{R}^2$.

2.2 Polyhedral and Zonotopic Sets

There exists a wealth of families which describe convex (or non-convex) sets with varying degrees of accuracy. An important limiting factor is the numerical reliability of their representation. That is, a particular family may be able to represent a great number of shapes but due to computationally expensive manipulations will be useless in practice. Usually there exists an inverse relation between flexibility of a family and the numerical cost of the representation.

2.2.1 Polyhedral Sets

Polyhedra[3] provide a useful geometrical representation for the linear constraints that appear in diverse fields such as linear control and optimization. In a convex setting, they provide a good compromise between complexity and flexibility. Due to their linear and convex nature, the basic set operations are relatively easy to implement [8]. Principally, this is related to their dual (half-spaces/vertices) representation [9] which allows to chose which formulation is best suited for a particular task. Note that the transformation from one representation to another may be time-consuming with various well-known algorithms: Fourier-Motzkin elimination—[10], Double Description method—[11], Equality Set Projection—[12]. With respect to their flexibility it is worthwhile to note that any convex body can be approximated arbitrarily well by a polytope [13].

---

[3]In here we will use the notions of *polyhedron* and *polytope*. The first represents the element of the polyhedral class under discussion whereas the latter denotes a bounded polyhedron.

The set operations implemented over the polyhedral family represent a main topic in the domain which lies at the intersection of convex geometry, mathematical programming and computer science [14]. To mention just a few, the algorithms used to implement the Minkowski addition (the sum over the space of sets), the Pontryagin difference (the non-dual counterpart of the sum over the space of sets—see formal definitions below), the translation between vertex and half-space representations are sensitive to the space dimension and the complexity of the chosen set representation (see [15, 16]).

We start by recalling some theoretical concepts (from Chap. 1 of [4]). Firstly, we provide the notion of $\mathcal{H}$-*polyhedron* which denotes an intersection of closed half-spaces:

Definition 2.3 A set $P \in \mathbb{R}^n$ is a $\mathcal{H}$-*polyhedron* if it can be implicitly presented in the form

$$P = \mathscr{P}(F, \theta) = \{x \in \mathbb{R}^n : Fx \le \theta\}, \tag{2.13}$$

for some $F \in \mathbb{R}^{m \times n}, \theta \in \mathbb{R}^m$.

The cone of a finite collection of vectors is defined by Definition 2.4 and the convex hull of a finite set of points by Definition 2.5:

Definition 2.4 For a finite collection of vectors $Y = \{y_1 \ldots y_d\} \subseteq \mathbb{R}^n$, the cone of Y is defined as

$$\text{Cone}(Y) \triangleq \{t_1 y_1 + \ldots t_d y_d : t_i \in \mathbb{R}_+\} = \{Yt, \ t \in \mathbb{R}_+^n\}.$$

Definition 2.5 For a finite collection of points $V = \{v_1 \ldots v_d\} \subseteq \mathbb{R}^n$, the convex hull of V is defined as

$$\text{Conv}(V) \triangleq \{\alpha_1 v_1 + \ldots \alpha_d v_d : \alpha_i \in \mathbb{R}_+, \sum_i \alpha_i = 1\} = \{V\alpha, \ \alpha \in \mathbb{R}_+^n, \mathbf{1}^T \alpha = 1\}.$$

In the context of set theory, the generalizations of addition (the *Minkowski sum*), difference (the *Pontryagin difference*) and distance (the *Hausdorff distance*) are provided below:

Definition 2.6 The Minkowski sum of two sets $P, Q \subseteq \mathbb{R}^n$ is defined to be

$$P \oplus Q = \{x + y : x \in P, y \in Q\},$$

and the Pontryagin difference is defined as

$$P \ominus Q = \{x \in P : x + y \in P, \forall y \in Q\}.$$

Definition 2.7 Given two convex sets P, Q, the Hausdorff distance is defined as

$$d_H(P, Q) = max\left\{\bar{d}_H(P, Q), \bar{d}_H(Q, P)\right\},$$

where $\bar{d}_H(P, Q) = \underset{x \in P}{max} \, \underset{y \in Q}{min} \, d(x, y)$, and $d(x, y)$ is a distance measured in a given norm in the $\mathbb{R}^n$ space.

Illustrative example for polyhedral set operations

In Fig. 2.4 polyhedral construction and some important set operations are illustrated. Figure 2.4a shows a bounded polyhedron (i.e., a polytope) which is characterized as half-space intersections or equivalently as convex sum of extremal vertices. Figure 2.4b delineates an unbounded polyhedron (a cone) which is again characterized by half-space intersections and cone representation (positive sum of rays). Lastly, in Fig. 2.4c the Minkowski addition and Pontryagin difference operations are illustrated graphically.

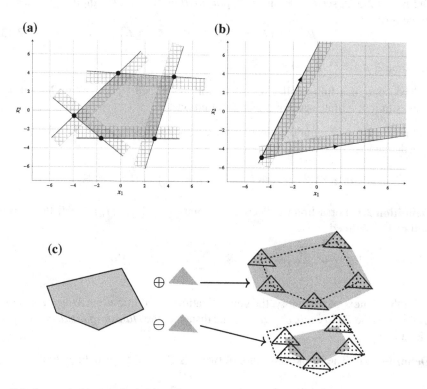

Fig. 2.4 Some primitives and operations for polytopic sets. **a** Convex hull of a polytope. **b** Cone. **c** Minkowski sum and Pontryagin difference representations

2.2.2 Zonotopic Sets

Zonotopes represent a particular class of polytopes (i.e., projections of a hypercube from a higher dimension) and can be described in "generator form" as

$$\mathscr{Z}(c, G) = \left\{ x \in \mathbb{R}^n : x = c + \sum_{i=1}^{m} \lambda_i g_i , |\lambda_i| \leq 1 \right\}, \tag{2.14}$$

with $i = 1 \ldots m$ and where $c \in \mathbb{R}^n$ represents the center and $G = \begin{bmatrix} g_1 \ldots g_m \end{bmatrix} \in \mathbb{R}^{n \times m}$ the matrix of generators. A zonotope defined as in (2.14) has some interesting properties [17]:

- is closed under linear transformations:

$$L\mathscr{Z}(c, G) = \mathscr{Z}(Lc, LG); \tag{2.15}$$

- is closed under the Minkowski sum:

$$\mathscr{Z}(c_1, G_1) \oplus \mathscr{Z}(c_2, G_2) = \mathscr{Z}(c_1 + c_2, \begin{bmatrix} G_1 & G_2 \end{bmatrix}). \tag{2.16}$$

Due to its particular structure, the number of vertices and facets for a zonotope (2.14) is significantly reduced in comparison with a randomly generated polyhedron.[4, 5]

In realistic situations, often the constraints are given in polytopic form but posses an encapsulated symmetry which allows a description in terms of zonotopic sets. Even when this is not the case, zonotopic approximations may be constructed. Since the generator representation (2.14) is more compact than either the half-space and vertex representations associated to polytopes, it becomes obvious why for numerical and theoretical reasons the zonotopes will be used whenever possible in the set constructions of this manuscript. However, we stress that this practical preference remains a subjective choice and the set-theoretic results appearing in the next developments hold for any class of sets (when convexity is mandatory, this requirement will be specified accordingly).

For polytopic sets, [20] proposes tight approximations in fixed directions and [21] discusses an iterative algorithm. A more general case is represented by convex bodies defined by nonlinear inequalities. Common characterizations of such sets include the

---

[4]From [18] we recall the following bounds on the number of facets f_i of order 'i' for a given zonope $\mathscr{Z}$ (bounds which are reached whenever the zonotope's generators are in general position):

$$f_0(\mathscr{Z}) \leq 2 \sum_{i=0}^{n-1} \binom{m-1}{i}, \quad f_{n-1}(\mathscr{Z}) \leq 2 \binom{m}{n-1}.$$

[5]A zonotope is topologically equivalent with an associated hyperplane arrangement (see Definition 2.1) offering thus efficient descriptions of the faces (e.g., by using reverse search algorithms as in [19]).

unit ball of the weighted p-norm (usually some weighted Euclidean norm defining an ellipsoid). In [22, 23] it is proven that any such Euclidean ball can be approximated arbitrarily close, in the sense of the Hausdorff distance, by a zonotope given by a uniform distribution on the surface of the (hyper)sphere.

2.3 Non-convex Region Description

In this manuscript we aim to characterize the complement of a given region which may or may not be convex. To this end let us consider a collection of (possibly overlapping) polyhedral sets in $\mathbb{R}^n$ that may be related to obstacles to better fix the ideas (usually encountered in $\mathbb{R}^2$ or $\mathbb{R}^3$ for motion planning problems):

$$\mathbb{S} = \bigcup_{l=1}^{N_o} S_l. \tag{2.17}$$

Describing the region of interest (2.17) as a union of polyhedral sets is not random: a polyhedral set is in fact a finite intersection of regions of type (2.2a)–(2.2b). We can then consider $\mathbb{H} = \{\mathscr{H}_i\}_{i \in \mathbb{I}}$ as the collection of all hyperplanes appearing in the description of polytopes S_l from (2.17) and construct the hyperplane arrangement (2.4).

Then, there exists a sign tuple σ_l which characterizes[6] S_l, i.e.,

$$S_l = \mathscr{A}(\sigma_l) = \bigcap_{i \in \mathbb{I}} \mathscr{R}_i^{\sigma_l(i)} = \mathscr{P}\left(\begin{bmatrix} \cdots \\ \sigma_l(i)h_i \\ \cdots \end{bmatrix}, \begin{bmatrix} \cdots \\ \sigma_l(i)k_i \\ \cdots \end{bmatrix}\right). \tag{2.18}$$

Further we can partition the collection of feasible tuples (2.5) into[7]:

(i) the collection of forbidden tuples

$$\Sigma^\bullet = \{\sigma \in \Sigma : \text{interior}\,(\mathscr{A}(\sigma) \cap \mathbb{S}) \neq \emptyset\}, \tag{2.19}$$

(ii) the collection of admissible tuples

$$\Sigma^\circ = \{\sigma \in \Sigma : \text{interior}\,(\mathscr{A}(\sigma) \cap \mathbb{S}) = \emptyset\} = \Sigma \setminus \Sigma^\bullet. \tag{2.20}$$

Remark 2.5 Note that we have chosen as classification criteria interior $(\mathscr{A}(\sigma) \cap \mathbb{S}) \neq \emptyset$ rather than $\mathscr{A}(\sigma) \subseteq \mathbb{S}$. The question is mote as long as the hypeplanes

---

[6]We assume without loss of generality that each set S_l is characterized by a unique tuple.

[7]The intersection is meant to discern the cells which overlap with the obstacles. The common frontier is irrelevant in this context and is formally discarded by taking the interior of the intersection into consideration.

generating the arrangement are taken from the half-space representation of the obstacles (in this case the inclusion and the intersection operations are equivalent). Still, we preferred the more robust interior $(\mathscr{A}(\sigma) \cap \mathbb{S}) \neq \emptyset$ which assignates a sign tuple as forbidden if there is any intersection between the current cell and the obstacle union.   ◆

With notation (2.19) and (2.20) we may describe regions $\mathbb{S}$ and $\overline{\mathbb{S}}$ in terms of forbidden and admissible tuples:

$$\mathbb{S} = \bigcup_{\sigma \in \Sigma^\bullet} \mathscr{A}(\sigma), \tag{2.21a}$$

$$\overline{\mathbb{S}} = \bigcup_{\sigma \in \Sigma^\circ} \mathscr{A}(\sigma). \tag{2.21b}$$

Note that $\Sigma^\bullet \cap \Sigma^\circ = \emptyset$ and $\Sigma^\bullet \cup \Sigma^\circ = \Sigma \subset \{-,+\}^N$.

Remark 2.6 At this point, we may ask what was first, the arrangement or the obstacles? The answer is that it depends on the problem at hand. Do we start with predefined obstacles (polyhedral sets) and find the hyperplane arrangement associated to them? Or, do we start with a predefined hyperplane arrangement (e.g., by grid-ing the space and assigning to the resulting cells admissible/forbidden values)? Each approach has its merits as on one hand we may have more precise bounds but difficult formulation and on the other hand we have over-approximations but under reduced/fixed complexity.   ◆

Illustrative example of non-convex regions

For the purpose of illustration let us consider a simple example as depicted in Fig. 2.5. A union of two obstacles, $\mathbb{S} = S_1 \cup S_2$ in $\mathbb{R}^2$ is considered. The forbidden regions are defined by 5 and respectively 3 hyperplanes (see Appendix A.4 for the numerical data). Furthermore, these partition the space into 37 cells from which 3 describe the obstacles and the rest characterize the feasible space $\overline{\mathbb{S}} = \mathbb{R}^2 \setminus S$. More precisely, $\Sigma^\bullet = \{\sigma_1, \sigma_2, \sigma_3\}$ is identified such that $S_1 = \mathscr{A}(\sigma_1)$ and $S_2 = \mathscr{A}(\sigma_2) \cup \mathscr{A}(\sigma_3)$ for $\sigma_1 = (+++-+++-)$, $\sigma_2 = (+-++++++)$ and $\sigma_3 = (--++++++)$. Note that the obstacle S_2 is described by more than one cell (this usually happens if a hyperplane from another obstacle cuts the obstacle under consideration, as for example the hyperplane $\mathscr{H}_1$ in Fig. 2.5). As it can be observed in the example, this is not an issue, it simply means that $\Sigma^\bullet$, the collection of forbidden tuples, has more tuples than there are obstacles.

2.3.1 Cell Merging

Recall that any of the cells from (2.21b) is described by a unique sign tuple and is disjoint with respect to the others (up to its boundary, see footnote 1). For our

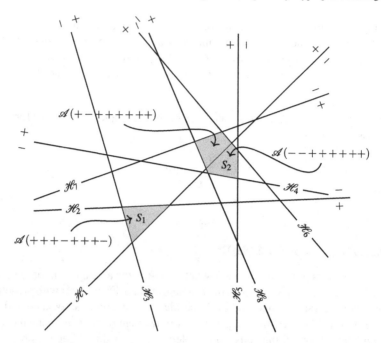

Fig. 2.5 Collection of obstacles and their associated hyperplane arrangement

purposes, we may be[8] satisfied with any collection of regions not necessarily disjoint which covers the feasible space. In this context, the question arises whether it is possible to merge the existing cells of (2.21b) into a reduced number of regions which still describe $\overline{\mathbb{S}}$. Note that this will help later on when it will lead to a compact mixed-integer representation.

In general, these *merged cells* have to be formed as unions of cells from (2.21b) such that the result is a convex set. Existing merging algorithms are usually computationally expensive because in general a union of convex sets is non-convex. In the present study the problem can be simplified by observing two properties of the cells from (2.21b) (and in general, from (2.4)):

- the sign tuples σ describe an adjacency graph since any two cells whose sign tuples differ at only one position are neighbors (i.e., they share a hyperplane as boundary),
- the union of any two adjacent cells is a polyhedra.

Using these properties a merged cell can be formally characterized.

Definition 2.8 The union of all cells $\mathscr{A}(\sigma)$ whose sign tuples retain the same values over a subset of indices ($\mathbf{i} \subset \mathbb{I}$) and span all possible combinations for the rest of

---

[8]In the forthcoming chapter constructions which accept both formulations will be discussed in detail.

indices ($\mathbb{I} \setminus \mathbf{i}$) is a "merged" cell. This cell is characterized by the sign tuple $\sigma^* \in \{-, *, +\}^N$ where $\sigma^*(i) \in \{-, +\}, \forall i \in \mathbf{i}$ and $\sigma^*(i) = *, \forall i \in \mathbb{I} \setminus \mathbf{i}$ such that:

$$\mathscr{A}(\sigma^*) = \bigcap_{\sigma^*(i) \neq '*', i \in \mathbb{I}} \mathscr{R}_i^{\sigma^*(i)} = \bigcup_{\substack{\sigma(i)=\sigma^*(i), \forall i \in \mathbf{i} \\ \sigma(j) \in \{-,+\}, \forall j \notin \mathbf{i}}} \mathscr{A}(\sigma). \tag{2.22}$$

$\blacklozenge$

Remark 2.7 Using Definition 2.8, the number of cells appearing in a merged cell is expected to be $2^{\mathrm{card}(\mathbb{I} \setminus \mathbf{i})}$ because "spanning all possible combinations" would lead to such a number. In practice the number of disjoint cells will be smaller since many of these sign combinations result in empty cells. $\blacklozenge$

With the previously given properties, merging algorithms can be employed (see, for example, [24], which adapt a "branch and bound" strategy). This is still cumbersome, and henceforth the problem is tackled in the Boolean algebra framework. The merging problem of regions from (2.21b) is functionally identical to the minimization of a Boolean function given in the "sum-of-products" form. A cell describing the (in)feasible cell from (2.3) corresponds to a "1" ("0") value in the truth-table at the position determined by its associated sign tuple, whereas infeasible sign tuples correspond to "don't care" values. It is then straightforward to apply minimization algorithms (Karnaugh maps, the Quine-McCluskey algorithm or the Espresso heuristic logic minimizer for example) in order to obtain Boolean minterms who describe the merged cells of (2.21b). Note that a similar merging procedure was proposed in [25] in order to deal with polyhedral piecewise affine systems.

The next theorem provides constructive means for generating merged cells (similar to the results in [26]).

Theorem 2.2 *Consider the forbidden and admissible regions* (2.21a), (2.21b) *characterized respectively by the sign tuples* (2.19) *and* (2.20). *Then, the feasible region* (2.21b) *is compactly described as a union of merged cells* (2.22):

$$\overline{\mathbb{S}} = \bigcup_{\sigma^*} \mathscr{A}(\sigma^*), \tag{2.23}$$

where $\sigma^* \in \{-, *, +\}^N$ *are given by the "sum-of-products" representation of the Boolean function* $f : \{-, +\}^N \to \{0, *, 1\}$ *verifying:*

$$f(\sigma) = 0, \ \forall \sigma \in \Sigma^\bullet, \tag{2.24a}$$
$$f(\sigma) = 1, \ \forall \sigma \in \Sigma^\circ, \tag{2.24b}$$
$$f(\sigma) = *, \ \forall \sigma \in \{-, +\}^N \setminus (\Sigma^\bullet \cup \Sigma^\circ). \tag{2.24c}$$

$\square$

Proof Let us consider function (2.24) and its truth-table: we have "0" for sign tuples characterizing cells inside (2.21a), "1" for sign tuples characterizing cells inside

(2.21b) and "don't care" values for all the sign tuples which correspond to empty cells (infeasible combinations of regions (2.2a)–(2.2b)). All that remains is to group the combinations which are "1" or "don't care" in the truth-table and express the function (2.24) in the canonical sum-of-products form. Each term of the product describes a region of form (2.22) thus reaching (2.23) and concluding the proof. ∎

Furthermore, Theorem 2.2 can be adapted so that we can actually avoid calculating the collection of feasible sign tuples (2.5). Rather, by using the set of cells which describe (2.21a) (usually containing many fewer cells than the total number, see Theorem 2.1) and Boolean algebra notions we provide a compact representation of (2.21b) without explicitly making the decomposition (2.4).

Corollary 2.1 *Consider the forbidden region* (2.21a) *characterized by the sign tuples* (2.19). *Then, the feasible region* (2.21b) *is compactly described as a union of merged cells of form* (2.22):

$$\overline{\mathbb{S}} = \bigcup_{\sigma^*} \mathscr{A}(\sigma^*), \tag{2.25}$$

where $\sigma^* \in \{-, *, +\}^N$ *are the sums from the "sum-of-products" representation of the Boolean function* $f : \{-, +\}^N \to \{0, 1\}$ *verifying*

$$f(\sigma) = 0, \ \forall \sigma \in \Sigma^\bullet, \tag{2.26a}$$
$$f(\sigma) = 1, \ \forall \sigma \in \{-, +\}^N \setminus \Sigma^\bullet. \tag{2.26b}$$

□

Proof Let us consider function (2.26) and its truth-table: we have "0" for sign combinations characterizing cells inside (2.21a) and "1" in all the other cases (regardless if the sign tuples describe admissible cells or infeasible combinations of regions). All that remains is to group the combinations which are "1" and express the function in the canonical sum-of-products form. Each term of the product describes a region of form (2.22) thus reaching (2.25). ∎

Several remarks are in order:

Remark 2.8 In the construction of the truth table of (2.26) as explained in the proof of Corollary 2.1 lies the major difference with respect to Theorem 2.2. By not computing the cells composing the feasible region, it is assumed implicitly (and conservatively) that all the remaining combinations of signs correspond to non-empty admissible cells (thus "1" in the table). This will result in a larger number of merged cells. More than that, one can even obtain merged cells which are infeasible (because the sign combination describes an empty region). Such a cell is denoted as feasible (and discarded otherwise) at a later post-processing stage, in contrast with the approach in Theorem 2.2 where the validation is made in a pre-processing stage.   ◆

Remark 2.9 For a compact representation it is suitable to write (2.24) or (2.26) in a minimal sum-of-products form. To this end, minimization algorithms (e.g., Karnaugh maps, the Quine-McCluskey algorithm or the Espresso heuristic logic minimizer) are employed. We emphasize on the latter technique which although heuristic gives near-optimal results and is orders of magnitude faster than other methods [27]. We note that a similar approach was proposed in [25] in order to deal with polyhedral piecewise affine systems. ♦

Remark 2.10 Note that a region (2.22) is described by at most $N - f$ hyperplanes, where f denotes the number of indices in the sign tuples which flip the sign. It makes sense then to, not only reduce the number of regions, but also to maximize the number of cells that go into the description of a region from (2.21b). ♦

Illustrative example of merging procedures

The example provided in Sect. 2.3 with the numerical data from Appendix A.4 is further used for exemplifying the results presented in the current section. The merging procedures given in Theorem 2.2 and Corollary 2.1 are applied in order to obtain *merged cells* as defined in Definition 2.8. In both cases the first step is to compute the Boolean function (2.24) ((2.26) respectively) and to simplify it to the canonical sum-of-products form.

For the small number of hyperplanes appearing in this example we may consider a graphical solution. That is, we represent the Boolean function as a Karnaugh map in Fig. 2.6 which can take the values '0'—obstacle cell, '1'—feasible cell and '*'— empty cell.

Reducing the map in Fig. 2.6 leads to the canonical sum-of-products form of (2.24):

$$f(\sigma) = \overline{\sigma}(2)\overline{\sigma}(8) + \overline{\sigma}(1)\overline{\sigma}(4) + \overline{\sigma}(6) + \overline{\sigma}(3) + \overline{\sigma}(7) + \overline{\sigma}(5).$$

$\sigma_1,\sigma_2,\sigma_3,\sigma_4$ \ $\sigma_5,\sigma_6,\sigma_7,\sigma_8$																
_ _ _ _	*	*	*	*	*	*	*	*	*	*	*	*	*	*	*	*
- - - +	*	*	*	*	*	*	*	*	*	*	1	*	*	*	*	*
- - + +	*	*	1	*	*	*	*	*	*	*	1	*	*	*	*	*
- - + -	*	*	*	*	*	*	*	*	*	*	*	*	*	*	*	*
- + - -	*	*	*	*	*	*	*	*	*	*	*	*	*	*	*	*
- + - +	*	*	*	*	*	*	*	*	*	*	*	*	*	*	*	*
- + + +	*	*	1	1	*	*	*	1	*	*	*	*	*	*	*	*
- + + -	*	*	*	*	*	*	*	1	*	*	*	*	*	*	*	*
+ - - -	*	*	*	*	*	*	*	*	*	*	*	*	*	*	*	*
+ - - +	*	*	*	*	*	*	*	*	*	*	1	*	*	*	*	*
+ - + +	*	*	1	*	*	*	*	*	*	*	1	*	*	*	*	*
+ - + -	*	*	*	*	*	*	*	*	*	*	*	*	*	*	*	*
+ + - -	*	*	*	*	*	*	*	*	1	*	1	1	*	*	*	*
+ + - +	*	*	*	*	*	*	*	*	*	*	1	*	*	*	*	*
+ + + +	*	*	0	1	*	*	*	1	*	*	0	*	*	*	*	*
+ + + -	*	*	1	1	1	*	*	1	1	*	1	1	1	*	*	0

Fig. 2.6 Karnaugh diagram for obtaining the reduced cell representation with Theorem 2.2

Each of the Boolean products appearing here denotes a merged cell. Taking them in order, we have: $\mathscr{A}(* - * * * * * -)$, $\mathscr{A}(- * * - * * * *)$, $\mathscr{A}(* * * * * - * *)$, $\mathscr{A}(* * - * * * * *)$, $\mathscr{A}(* * * * * * -*)$ and $\mathscr{A}(* * * * - * * *)$. The union of these merged cells describes the feasible region (2.21b) as in (2.25).

For illustration, Fig. 2.7 delineates the merged cell $\mathscr{A}(* - * * * * * -) = \mathscr{R}_2^- \cap \mathscr{R}_8^-$ and the hyperplane arrangement which spanned them.

Note that, in addition to reducing the number of regions in (2.25) comparative with (2.21b)—from 24 to 6, we have also reduced the number of hyperplanes appearing in the region's half-space representation (see Remark 2.10). These differences become more significant with a larger number of hyperplanes and of obstacles.

The same reasoning, but under the conservative view expressed in Corollary 2.1, can be applied to the problem. In the truth-table depicted in Fig. 2.8 we assign '0' whenever the cell corresponds to an obstacle and '1' otherwise. The result, once the associated Boolean function (2.26) is put into the canonical sum-of-products form:

$$f(\sigma) = \overline{\sigma}(1)\overline{\sigma}(8) + \overline{\sigma}(2)\overline{\sigma}(8) + \overline{\sigma}(4)\sigma(8) + \sigma(2)\sigma(4)$$
$$+ \overline{\sigma}(7) + \overline{\sigma}(6) + \overline{\sigma}(5) + \overline{\sigma}(3),$$

are the merged cells $\mathscr{A}(- * * * * * * *)$, $\mathscr{A}(* - * + * * * *)$, $\mathscr{A}(* * * * * * * -)$, $\mathscr{A}(* * - + * * * *)$, $\mathscr{A}(* * * * * * -*)$, $\mathscr{A}(* * * * * - * *)$, $\mathscr{A}(* * * * - * * *)$ and $\mathscr{A}(- - + * * * * *)$. As before, the union of these merged cells describes the fea-

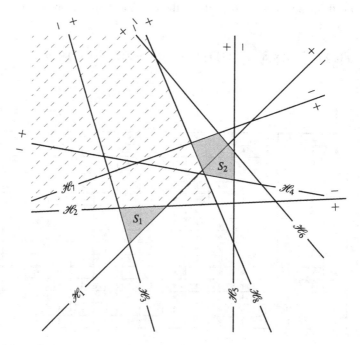

Fig. 2.7 Hyperplane arrangement with merged cells computed as in Theorem 2.2

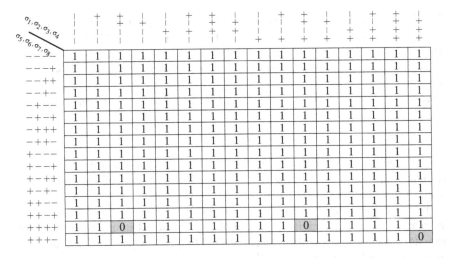

Fig. 2.8 Karnaugh diagram for obtaining the reduced cell representation with Corollary 2.1

sible region (2.21b) as in (2.25). In this particular case we observe that all the sign tuples correspond to non-empty regions (which, in general might not be the case— Remark 2.8). The drawback is that, even for this relatively small example the implicit representation is more conservative than the explicit one (8 merged cells instead of 6).

2.3.2 Numerical Considerations

The two main issues discussed in this chapter are the computation of the hyperplane arrangement (2.3) and the merging procedures which lead to compact representation (2.22). The first issue is an enumeration problem. The second issue, at least in the approach followed here, reduces to the minimization of a Boolean function. Both these problems have worst-case resolution times of exponential complexity. In practice there are algorithms which provide (sub)optimal solutions in a reasonable time. These algorithms require a broad mathematical background and fall beyond the scope of this book. Rather, we will consider several representative examples (with the numerical data provided in Appendices A.3–A.9) and apply existing algorithms to observe the computation times and various properties of interest.

The computation times are of course dependent on the hardware and software platforms and therefore the numbers obtained should not be seen as "best" values, rather they should serve to underline the link between computational difficulty and various design parameters. The numbers illustrated in the following tables have been obtained after extensive simulation in Matlab 2014a, using MPT3 toolbox [28] and the Espresso heuristic logic minimizer [29, 30]. MPT contains routines which permit the computation of hyperplane arrangements (used internally for simplifying

Table 2.1 Computation times and number of feasible cells for hyperplane arrangements in $\mathbb{R}^2$

(N, d)	$(3, 2)$	$(8, 2)$	$(10, 2)$	$(15, 2)$	$(20, 2)$	$(25, 2)$	$(30, 2)$
Card Σ	7	27	46	98	179	252	365
Card Σ cf. (2.8)	7	56	90	210	380	600	870
Time [s]	0.7314	1.2375	1.9720	3.8619	7.7412	12.7355	17.8422

piecewise affine representations [25, 31]) and the Espresso minimizer produces suboptimal simplifications of Boolean functions which are order of magnitude faster than optimal algorithms [32].

Table 2.1 illustrates the times required for the computation of a hyperplane arrangement, starting from its collection of hyperplanes. The first row enumerates various collection of hyperplanes characterized through the pair (N, d)—number of hyperplanes and space dimension. The second and third rows provide the number of cells obtained with the MPT3 routines and as the upper bound given in (2.8). Note that the computed number of cells is usually smaller. This is due to the fact that the arrangement might not be in general position and because the MPT3 routine used counts only the cells inside a finite box (a reasonable assumption for realistic control problems—we took here a bound $[-20, 20]$ along eachs axis). Finally, the last row gives the computation times. As it can be seen, the values obtained are reasonable, at least for the $\mathbb{R}^2$ case which can cope with robotic applications for example.

In the following, the proposed merging procedures (see Theorem 2.2 and Corollary 2.1) are analyzed and solved via the Espresso minimizer. The data sent to the minimizer is divided into three disjoint categories: combinations of signs to which corresponds '1'—an admissible cell; combinations to which corresponds '0'—a forbidden cell and combinations to which corresponds '*'—empty cells. Furthermore, the input can be composed from any combination of these categories. Two of these cases hold interest to us:

(i) both the '0' and '1' cases are given explicitly; this corresponds to Theorem 2.2,
(ii) only the '0' cases are given explicitly; this corresponds to Corollary 2.1.

Table 2.2 gives the arrangements corresponding to the examples in Table 2.1 and the number of forbidden tuples in the first two rows. The next four rows are paired two by two and provide the number of merged cells and the time to compute them in case (i) and (ii) respectively (card $\Sigma^{\bullet,1}$ is the number of forbidden tuples, card $\Sigma^*_{(i)/(ii)}$ is the number of merged cells and time$_{(i)/(ii)}$ are the computation times).

Several observations can be made. First, the explicit approach is clearly better in terms of number of merged cells: in each case the number obtained through the implicit method is larger (sometimes much larger). Interestingly the computation times favor the explicit approaches in most all instances. Here we have to note that the explicit approach requires the computation of the arrangement as a pre-processing step. Including this additional time as well the implicit approach becomes clearly faster than the explicit one. A silver lining is the observation that the arrangement

Table 2.2 Computation times and number of merged cells for hyperplane arrangements in $\mathbb{R}^2$

(N, d)	$(3, 2)$	$(8, 2)$	$(10, 2)$	$(15, 2)$	$(20, 2)$	$(25, 2)$	$(30, 2)$
Card $\Sigma^{\bullet,1}$	1	3	4	9	9	31	22
Card $\Sigma^*_{(i)}$	3	6	8	11	10	19	16
Time$_{(i)}$ [s]	0.1700	0.1061	0.1141	0.1401	0.1665	0.2632	0.3612
Card $\Sigma^*_{(ii)}$	3	8	11	16	23	45	43
Time$_{(ii)}$ [s]	0.1561	0.1022	0.1113	0.1423	0.1766	0.3759	0.4665

Table 2.3 Computation times and number of merged cells for hyperplane arrangements in $\mathbb{R}^2$

(N, d)	$(3, 2)$	$(8, 2)$	$(10, 2)$	$(15, 2)$	$(20, 2)$	$(25, 2)$	$(30, 2)$
Card $\Sigma^{\bullet,2}$	2	5	7	23	14	40	37
Card $\Sigma^*_{(i)}$	2	7	7	10	11	20	21
Time$_{(i)}$ [s]	0.1670	0.1071	0.1062	0.1554	0.1787	0.2580	0.4630
Card $\Sigma^*_{(ii)}$	2	8	13	24	27	51	55
Time$_{(ii)}$ [s]	0.1687	0.1232	0.1312	0.1616	0.1944	0.6450	1.0896

needs to be computed only once. After this operation, the obstacle classification reduces to a separation between admissible and forbidden tuples.

While the number of hyperplanes and the space dimension have a clear influence on the computation time not the same can be said about the forbidden tuples. Even keeping constant the number of hyperplanes, it is intuitive that changing the distribution and number of forbidden cells will modify the number and shape of the merged cells. To illustrate the point, in Table 2.3 we perform the same operations as in Table 2.2 but for a different collection of forbidden tuples, $\Sigma^{\bullet,2}$, these and the initial forbidden tuples are available in Appendix A.

As expected, the number of merged cells is larger (although there is no formal relation between this number and the number of forbidden tuples). The computation times are in almost all cases larger (again, due to the increased complexity of the problem).

To have a better grasp of the computation times for the merging procedures under the various methods and initial data, Fig. 2.9 plots the relevant results provided in Tables 2.2 and 2.3.

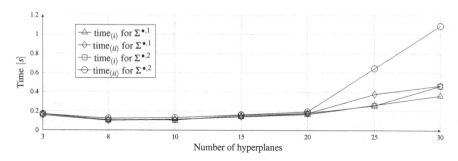

Fig. 2.9 Illustration of the merging procedure computation times provided in Tables 2.2 and 2.3

2.4 Notes and Comments

The scope of this chapter was to present basic combinatorial and set-theoretic notions necessary to characterize the feasible space (e.g., the complement of a union of obstacles). Assuming polyhedral characterization of the forbidden regions we can take the hyperplanes defining their borders and use them to create an appropriate hyperplane arrangement. From our viewpoint, the main advantage is represented by the fact that this arrangement characterizes each cell in terms on which side (either negative or positive) of the hyperplanes it lies. Hence, we no longer need to provide a geometrical description of the cell but rather of the tuple of signs which uniquely defines it. Furthermore, to characterize each of these cells as forbidden (i.e., the part of the space defining the obstacles) or feasible (i.e., the remaining part) it suffices to gather and separate the sign tuples into two disjoint collections. Moreover, since we are equally satisfied with overlapping descriptions of the feasible space we also present the notion of merged cells (hence reducing the complexity of the feasible space description).

Nevertheless, while here we presented some of the basics, extensive additional information can be gathered from a multitude of sources. Classical monographs about polyhedral or zonotopic notions can be found in [4, 9, 33] whereas, their application in control are detailed in [34, 35], amongst others. Next, detailed mathematical descriptions and various applications of hyperplane arrangements notions are provided by Zaslavsky, Orlik, Geyer, De Concini, Procesi and others in [1, 6, 24, 36–38].

References

1. Orlik, P.: Hyperplane arrangements. In: Floudas, C., Pardalos, P. (eds.) Encyclopedia of Optimization, pp. 1545–1547. Springer, US (2009)
2. Stanley, R.: An introduction to hyperplane arrangements. In: Lecture notes, IAS/Park City Mathematics Institute. Citeseer (2004)
3. Blanchini, F.: Set invariance in control-a survey. Automatica **35**(11), 1747–1767 (1999)

4. Ziegler, G.: Lectures on Polytopes, vol. 152. Springer (1995)
5. Birkhoff, G.: Abstract linear dependence and lattices. Am. J. Math. 800–804 (1935)
6. Zaslavsky, T.: Facing up to arrangements: face-count formulas for partitions of space by hyperplanes. Am. Math. Soc. (1975)
7. Buck, R.: Partition of space. Am. Math. Monthly 541–544 (1943)
8. Loechner, V.: Polylib: a library for manipulating parameterized polyhedra (1999)
9. Motzkin, T., Raiffa, H., Thompson, G., Thrall, R.: The double description method. Contrib. Theory Games **2**, 51 (1959)
10. Dantzig, G.: Fourier-Motzkin elimination and its dual. Technical Report DTIC Document (1972)
11. Fukuda, K.: CDD/CDD+ Reference Manual. Institute for operations Research ETH-Zentrum, Zurich (1999)
12. Jones, C.N., Kerrigan, E.C., Maciejowski, J.M.: A new algorithm for the projection of polytopes in halfspace representation. Citeseer (2004)
13. Bronstein, E.: Approximation of convex sets by polytopes. J. Math. Sci. **153**(6), 727–762 (2008)
14. Wilde, D.: A library for doing polyhedral operations. Int. J. Parallel, Emergent Distrib. Syst. **15**(3), 137–166 (2000)
15. Gritzmann, P., Klee, V.: On the complexity of some basic problems in computational convexity: I. containment problems. Discrete Math. **136**(1–3), 129–174 (1994)
16. Gritzmann, P., Klee, V.: On the complexity of some basic problems in computational convexity: II. volume and mixed volumes. NATO ASI Ser. C Math. Phys. Sci. Adv. Study Inst. **440**, 373–466 (1994)
17. Fukuda, K.: From the zonotope construction to the Minkowski addition of convex polytopes. J. Symbolic Comput. **38**(4), 1261–1272 (2004)
18. Fukuda, K.: Polytope examples. ftp://ftp.ifor.math.ethz.ch/pub/fukuda/reports/polyfaq041121.pdf
19. Avis, D., Fukuda, K.: Reverse search for enumeration. Discrete Appl. Math. **65**(1), 21–46 (1996)
20. Alamo, T., Bravo, J., Camacho, E.: Guaranteed state estimation by zonotopes. Automatica **41**(6), 1035–1043 (2005)
21. Dang, T.: Approximate reachability computation for polynomial systems. Hybrid Syst. Comput. Control 138–152 (2006)
22. Bourgain, J., Lindenstrauss, J.: Distribution of points on spheres and approximation by zonotopes. Isr. J. Math. **64**(1), 25–31 (1988)
23. Linhart, J.: Approximation of a ball by zonotopes using uniform distribution on the sphere. Archiv der Mathematik **53**(1), 82–86 (1989)
24. Geyer, T., Torrisi, F., Morari, M.: Optimal complexity reduction of piecewise affine models based on hyperplane arrangements. In: Proceedings of the 23th American Control Conference, vol. 2, pp. 1190–1195. Boston (2004)
25. Geyer, T., Torrisi, F., Morari, M.: Optimal complexity reduction of polyhedral piecewise affine systems. Automatica **44**(7), 1728–1740 (2008)
26. Prodan, I., Stoican, F., Olaru, S., Niculescu, S.I.: Enhancements on the hyperplanes arrangements in mixed-integer techniques. J. Optim. Theory Appl. **154**(2), 549–572 (2012)
27. Brayton, R., Hachtel, G., McMullen, C., Sangiovanni-Vincentelli, A.: Logic minimization algorithms for VLSI synthesis, vol. 2. Springer (1984)
28. Herceg, M., Kvasnica, M., Jones, C., Morari, M.: Multi-parametric toolbox 3.0. In: Proceedings of the European Control Conference, pp. 502–510. Zürich (2013). http://control.ee.ethz.ch/~mpt
29. McGeer, P., Sanghavi, J., Brayton, R., Sangiovanni-Vincentelli, A.: Espresso-signature: a new exact minimizer for logic functions. IEEE Trans. Very Large Scale Integr. (VLSI) Syst. **1**(4), 432–440 (1993)
30. Hlavička, J., Fišer, P.: Boom: a heuristic boolean minimizer. In: Proceedings of the 2001 IEEE/ACM International Conference On Computer-aided Design, pp. 439–442. IEEE Press (2001)

31. Geyer, T., Torrisi, F., Morari, M.: Efficient mode enumeration of compositional hybrid systems. Int. J. Control **83**(2), 313–329 (2010)
32. Kunz, W., Stoffel, D.: Logic optimization. In: Reasoning in Boolean Networks, pp. 101–161. Springer (1997)
33. Schneider, R.: Convex Bodies: the Brunn-Minkowski Theory. Cambridge University Press (1993)
34. Althoff, M., Stursberg, O., Buss, M.: Computing reachable sets of hybrid systems using a combination of zonotopes and polytopes. Nonlinear Anal. Hybrid Syst. **4**(2), 233–249 (2010)
35. Blanchini, F., Miani, S.: Set-Theoretic Methods In Control. Birkhauser (2007)
36. Edelsbrunner, H., Seidel, R., Sharir, M.: On the zone theorem for hyperplane arrangements. New Results New Trends Comput. Sci. 108–123 (1991)
37. Orlik, P., Terao, H.: Arrangements of Hyperplanes, vol. 300. Springer (1992)
38. De Concini, C., Procesi, C.: Topics in Hyperplane Arrangements, Polytopes and Box-Splines. Springer (2010)

Chapter 3
Mixed-Integer Representations

As already mentioned, optimization problems over non-convex regions are a well established topic. Mixed-integer formulations provide one of the best ways of dealing with these problems [1], at least from the point of view of the feasible domain representation. The advantage consists in reformulating the problem into a mixed continuous and binary form which is manageable for reasonable dimensions and which allows to use efficient algorithms. However, the computational complexity is highly dependent on the size of the binary part which limits its usefulness to relatively small-size problems. There are works that try reducing the number of binary variables used in the problem formulation. For example, in [2] and the references therein, a logarithmic formulation is discussed. This and other techniques help to reduce the computational burden but, ultimately, the complexity of the problem is directly related with the difficulty of representing the feasible region. Hence, using the notions and results introduced in Chap. 2, the present chapter shows various constructions with the aim of bringing the non-convex regions in a compact mixed-integer representation. We do not select a single construction but rather propose several equivalent ones. This is done mainly because the complexity of the mixed-integer formulation is highly dependent on the geometry of the problem (i.e., number, shape and distribution of the forbidden regions). In particular, we propose to describe the region of interest from three different angles: in terms of its obstacles; through the polyhedral decomposition of the feasible region and, lastly, by explicitly using the hyperplane arrangement in the mixed-integer description.

3.1 Classical Formulation

Before fully stepping into the general case of a union of polyhedral obstacles as presented in Chap. 2, see (2.4)–(2.17), we present first the simplest obstacle avoidance problem and its mixed-integer representation.

© The Author(s) 2016

I. Prodan et al., *Mixed-Integer Representations in Control Design*,
SpringerBriefs in Control, Automation and Robotics,
DOI 10.1007/978-3-319-26995-5_3

Let us consider a polytope $S \subset \mathbb{R}^n$ as a bounded intersection of half-spaces[1]:

$$S = \bigcap_{i \in \mathbb{I}} \mathscr{R}_i^+, \tag{3.1}$$

and its complement $\overline{S}$ over $\mathbb{R}^n$ as the union of regions that cover all space except S:

$$\overline{S} = \overline{\bigcap_{i \in \mathbb{I}} \mathscr{R}_i^+} = \bigcup_{i \in \mathbb{I}} \overline{\mathscr{R}_i^+} = \bigcup_{i \in \mathbb{I}} \mathscr{R}_i^-. \tag{3.2}$$

For further use let us recapitulate some standard notions involving sets and their complements (used in (3.2) and thereafter). That is, let there be two sets X and Y, then the following properties hold:

- $\overline{X \cap Y} = \overline{X} \cup \overline{Y}$;
- $\overline{X \cup Y} = \overline{X} \cap \overline{Y}$;
- $\overline{Y \backslash X} = X \cup \overline{Y}$.

As explained in the introduction, it is often the case that the feasible region of an optimization problem is non-convex and is actually the complement of some convex set. With the present notation, this feasible region is (3.2) and in order to reach a tractable formulation, we have to use mixed-integer techniques. By adding the binary variables $(\alpha_1 \dots \alpha_N) \in \{0, 1\}^N$ the polytopic set in the extended space of *state* and *auxiliary binary variables* is obtained:

$$-h_i x \leq -k_i + M\alpha_i, \ i \in \mathbb{I}, \tag{3.3a}$$

$$\sum_{i \in \mathbb{I}} \alpha_i \leq N - 1, \tag{3.3b}$$

with M a constant chosen appropriately (that is, significantly larger than the rest of the variables and playing the role of a relaxation constant—hence, the "big-M" name for this type of mixed-integer formulation [2]).

Remark 3.1 Constraints (3.3a)–(3.3b) can be reduced to the original feasible region (3.2) by suitable choices of the binary variables. For example, a region $\mathscr{R}_i^-$ can be obtained from (3.3a) by setting the associated binary variable to be "0":

$$(\alpha_1 \dots \alpha_N) = (1 \dots 1 \underbrace{0}_{i} 1 \dots, 1). \tag{3.4}$$

Note that the converse is false since no choice of binary variables can lead to the description of a region $\mathscr{R}_i^+$. If a binary variable is "1", the corresponding inequality degenerates such that it covers any point $x \in \mathbb{R}^n$ (this represents the limit case for

---

[1] The "+" superscript was chosen for the homogeneity of notation, equivalently one could have chosen any combination of signs from (2.2a)–(2.2b) in order to describe the polytope S.

$M \to \infty$). The condition (3.3b) is thus required to ensure that at least one binary value is "0" and, consequently, that at least one inequality is verified. ◆

Remark 3.2 Usually, the issue of choosing a "sufficiently large value" for the constant M remains vague (numerically, M has to be finite, hence the constraints (3.3a) do not 'disappear' for $\alpha_i = 1$ but rather should become 'insignificant' with respect to the feasible space). For a finite bounding set around the region of interest (say $\mathbb{X}$) there always exists a finite M obtained via a LP problem which makes the half-spaces redundant (3.3a) with respect to $\mathbb{X}$:

$$M = \max_{i, M_i \geq 0} \left(\max_{x \in \mathbb{X}} \{M_i = k_i - h_i x\} \right). \tag{3.5}$$

Note that instead of taking a single M which makes all the half-spaces redundant, we may consider elementwise M_i which make redundant the i-th half-space. ◆

As it can be seen in the representation (3.3a)–(3.3b), a binary variable is associated to each region of form (2.2b) in the description of the feasible region (3.2). Obviously, this may increase the number of binary variables significantly. A commonly used solution is the *"logarithmic formulation"* of the binary part.

Let us recall and adapt Proposition 3.1 from [3].

Proposition 3.1 *For each region $\mathscr{R}_i^-$ from (3.2) a unique combination of binary variables $\lambda^i \in \{0, 1\}^{\lceil log_2 N \rceil}$ is associated. Then, the affine functions $\alpha_i : \{0, 1\}^{\lceil log_2 N \rceil} \to \{0\} \cup [1, \infty)$ can be constructed:*

$$\alpha_i(\lambda) = \sum_{j=0}^{\lceil log_2 N \rceil} \left(\lambda_j^i + (1 - 2\lambda_j^i) \cdot \lambda_j \right), \tag{3.6}$$

where index 'j' denotes the jth variable and λ_j^i its value for the tuple λ^i, associated to region $\mathscr{R}_i^-$.

Proof See the details in [3]. ∎

3.2 Analysis of the Unallocated Tuples

While the logarithmic approach allows a more compact representation (for N components characterizing a region, $N_0 = \lceil log_2 N \rceil$ binary variables will suffice) it also raises issues with the mixed-integer representation. In the previous section we have associated to each region of interest a tuple of binary variables but usually some remain unassociated and still feasible. If such a tuple would be selected by the solver, it would lead to a degenerate solution where the entire space is feasible. Hence, these unallocated tuples have to be made infeasible such that they can never be selected (similarly to condition (3.3b)).

Hereinafter, two methods will be presented, one based on the geometric interpretation of the binary variables associated to the original problem and one based on a specific ordering of the selected tuples.

3.2.1   Geometric Approach

The tuples left unallocated will be labeled as *prohibited*, and additional constraint inequalities will have to be added to the extended set of constraints (3.3a) in order to effectively render them infeasible to an optimization routine. These restrictions are justified by the fact that, under construction (3.6), an unallocated tuple will not enforce the verification of any of the constraints of (3.3a) (see Remark 3.1). It then becomes evident that the single constraint of (3.3b) has to be substituted by a set of constraints that explicitly make all the unallocated tuples infeasible.

The next corollary of Proposition 3.1 provides the means to construct an inequality which renders a tuple infeasible.

Corollary 3.1 *[3] Let there be a tuple $\lambda^i \in \{0, 1\}^{N_0}$, then the constraint*

$$\sum_{j=1}^{N_0} \left(\lambda^i_j + (1 - 2\lambda^i_j) \cdot \lambda_j \right) \geq \varepsilon, \tag{3.7}$$

with $\varepsilon \in (0, 1)$ makes λ^i, and only it, infeasible.   □

Proof The left side of the inequality (3.7) will vanish only at tuple λ^i, and for the rest of the tuples in the discrete set $\{0, 1\}^{N_0}$ will give values greater than or equal to 1. Thus, the only tuple made infeasible by inequality (3.7) is λ^i.   ∎

The number of unallocated tuples, N_{int}, may be significant, a relationship in this sense being given by:

$$0 \leq N_{int} \leq 2^{\lceil log_2 N \rceil} - 2^{\lceil log_2 N \rceil - 1} - 1 = 2^{\lceil log_2 N \rceil - 1} - 1, \tag{3.8}$$

with the bound reached for the most unfavorable case of $N = 2^{\lceil log_2 N \rceil - 1} + 1$. Consequently, by associating to each of the unallocated tuples an inequality as in Corollary 3.1 we will have to deal with a large number of inequalities (3.7).

The number of constraints can be reduced by noting (as previously mentioned) that the association between regions and tuples is arbitrary. One could then choose favorable associations which will permit more than one tuple to be removed through a single inequality. To this end, the next proposition is introduced:

Proposition 3.2 *[3] Let there be a collection of tuples $\{\lambda^i\} \subset \{0, 1\}^{N_0}$, which completely spans a d-facet[2] of hypercube $\{0, 1\}^{N_0}$. Let **i** be the set of the $N_0 - d$ indices,*

---

[2]Here, d denotes the degree of the facet, ranging from 0 for extreme points to $N_0 - 1$ for faces of the hypercube.

which retain a constant value over all the tuples $\{\lambda^i\}$ composing the facet. Then, the inequality

$$\sum_{j \in i} \left(\lambda_j^i + (1 - 2\lambda_j^i) \cdot \lambda_j \right) \geq \varepsilon, \tag{3.9}$$

with $\varepsilon \in (0, 1)$ makes all λ^i, and only them, infeasible.   □

Proof Geometrically, the tuples are extreme points on the hypercube $\{0, 1\}^{N_0}$ and the constraints are represented as linear inequalities which separate the points of the hypercube via half-spaces. If a set of tuples completely spans a d-facet, it is always possible to isolate a half-space that separates the points of the d-facet from the rest of the hypercube.   ■

By a suitable association between feasible cells and tuples, we may label as unallocated the extreme points which compose entire facets on the hypercube $\{0, 1\}^{N_0}$, which permits to apply Proposition 3.2 in order to obtain the constraints (3.9).

Remark 3.3 By writing N_{int} as a sum of consecutive powers of 2, i.e., $N_{int} = \sum_{i=0}^{\lceil \log_2 N_{int} \rceil - 1} b_i 2^i$, an upper bound N_{hyp} for the number of inequalities (3.9) can be computed:

$$N_{hyp} = \sum_{i=0}^{\lceil \log_2 N_{int} \rceil - 1} b_i \leq \lceil \log_2 N_{int} \rceil, \tag{3.10}$$

where $b_i \in \{0, 1\}$.   ◆

Remark 3.4 Note that (3.10) offers an upper bound for the number of inequalities, but practically the minimal value can be improved depending on the method used for constructing the separating hyperplanes and on the partitioning of the tuples between the allocated and unallocated subsets.   ◆

3.2.2 Algebraic Approach

Another approach which assumes a particular tuple allocation is explored in [4]. The authors show that as long as the sign tuples are ordered monotonously (such that the collection of tuples can always be separated into two contiguous intervals) a single inequality will always suffice to separate between the allocated and unallocated tuples.

With the present notation, Theorem 2.1 of [4] becomes:

Proposition 3.3 *Assuming that the tuples are allocated monotonously ascending, i.e., to the i-th region corresponds $\lambda^i = (b_0^i \dots b_{N_0-1}^i)$ where $\lambda_j^i = b_j^i$ is the j-th term* *in the binary decomposition of number* $i = \sum_{j=0}^{N_0-1} 2^{N_0-j-1} b_j^i$ *and that the last allocated*

tuple corresponds to $\bar{i}$, *the inequality*

$$\sum_{j=0}^{N_0-1} 2^{N_0-j-1} \lambda_j \le \bar{i} + \varepsilon, \tag{3.11}$$

with $\varepsilon \in (0, 1)$ *makes all* λ^i *with* $i > \bar{i}$, *and only them, infeasible.* □

Proof The left side of (3.11), $\sum_{j=0}^{N_0-1} 2^{N_0-j-1} \lambda_j$, is actually computing the number codified in the binary representation $(\lambda_0 \dots \lambda_{N_0-1})$. Since we assumed that the tuples corresponding to $i \in \{0 \dots \bar{i}\}$ are allocated, the unallocated tuples correspond to $i \in \{\bar{i}+1 \dots 2^{N_0}\}$ and it suffices to have a constraint which tests if the codified number is larger or smaller than $\bar{i}$. ■

Remark 3.5 Proposition 3.3 assumes that the allocated tuples are those corresponding to the sequence $\{0 \dots \bar{i}\}$. Similarly, the last $\bar{i} + 1$ numbers can be allocated, i.e., the sequence $i \in \{2^{N_0} - \bar{i} \dots 2^{N_0}\}$. Hence, the constraint (3.11) becomes:

$$\sum_{j=0}^{N_0-1} 2^{N_0-j-1} \lambda_j \ge 2^{N_0} - \bar{i} - 1 + \varepsilon, \tag{3.12}$$

with $\varepsilon \in (0, 1)$ a scalar. ◆

Illustrative example for the tuple allocation

For a good understanding of the previously discussed formulations let us consider a simple $\mathbb{R}^2$ illustration. We develop here on the numerical example presented in Chap. 2 where a collection of obstacles and their associated hyperplane arrangement were presented (see also Fig. 2.5 with the numerical data provided in Appendix A.4).

First, we consider the obstacle S_1 from Fig. 2.5 represented as in (3.1) (i.e., via an intersection of half-spaces of the form $\{h_i x \le k_i\}$ where $i = 1, 2, 3$). Then, the mixed-integer formulation can be written as described in (3.3a)–(3.3b):

$$-h_1 x \le -k_1 + M\alpha_1, \tag{3.13a}$$
$$-h_2 x \le -k_2 + M\alpha_2, \tag{3.13b}$$
$$-h_3 x \le -k_3 + M\alpha_3, \tag{3.13c}$$
$$\alpha_1 + \alpha_2 + \alpha_3 \le 2. \tag{3.13d}$$

As it can be observed, 3 binary variables are needed, one for each constraint, and an additional constraint, (3.13d), to force at least one of the regions (2.2b) to be active at any time. An alternative construction is the one provided in Proposition 3.1 with only $\lceil log_2 3 \rceil = 2$ binary variables:

$$- h_1 x \leq -k_1 + M(\lambda_1 + \lambda_2), \tag{3.14a}$$
$$- h_2 x \leq -k_2 + M(1 - \lambda_1 + \lambda_2), \tag{3.14b}$$
$$- h_3 x \leq -k_3 + M(1 + \lambda_1 - \lambda_2), \tag{3.14c}$$
$$2 - \lambda_1 - \lambda_2 > 0. \tag{3.14d}$$

It is worth underlining that a constraint is needed in order to make the problem well-posed. Constraint (3.14d) makes $(\lambda_1, \lambda_2) = (1, 1)$ infeasible and thus, ensures that at least one of the constraints (3.14a)–(3.14c) is active. As discussed in Sect. 3.2 there are several ways of writing the constraints which make infeasible the unallocated tuples. In this particular case, only one tuple remains unallocated, $(1, 1)$ and hence any method will do. In particular, using Proposition 3.1 it can be observed that (3.14d) interdicts tuple $(1, 1)$ and only it.

Figure 3.1 highlights region $\mathcal{R}_2^-$ and shows to which combination of binary variables it corresponds in either formulation (3.13) or (3.14). For the first case, $\mathcal{R}_2^-$ is obtained by taking $(\alpha_1, \alpha_2, \alpha_3) = (1, 1, 0)$ whereas, for the second case, we have $(\lambda_1, \lambda_2) = (1, 0)$.

Note that in both classical and logarithmic formulation the mapping between regions and sign tuples is not unique. The only constraint is that the relation needs to be bijective. To keep it simple, hereinafter, the sign tuples are mapped in lexicographical order, i.e., the 'i'-th region is codified under the logarithmic formulation

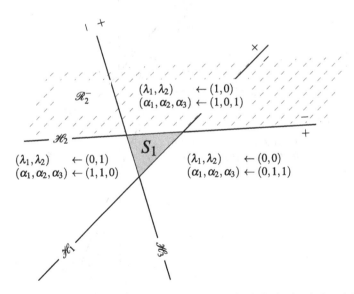

Fig. 3.1 Illustration of mixed-integer formulation for (3.2) in both classical and logarithmic formulation

by the same tuple corresponding to the binary representation of 'i' (also useful from the point of view of Proposition 3.3).

Second, let us consider the obstacle S_2 in $\mathbb{R}^2$ from Fig. 2.5. Keeping the notation from Appendix A.4, this obstacle is defined by the intersection of half-spaces $\{h_i x \leq k_i\}$ where $i = 4, 5, 6, 7$. Similarly with the previous case we write first the classical mixed-integer formulation as in (3.3a)–(3.3b), where 5 binary variables are needed:

$$-h_4 x \leq -k_4 + M\alpha_4, \tag{3.15a}$$
$$-h_5 x \leq -k_5 + M\alpha_5, \tag{3.15b}$$
$$-h_6 x \leq -k_6 + M\alpha_6, \tag{3.15c}$$
$$-h_7 x \leq -k_7 + M\alpha_7, \tag{3.15d}$$
$$-h_8 x \leq -k_8 + M\alpha_8, \tag{3.15e}$$
$$\alpha_4 + \alpha_5 + \alpha_6 + \alpha_7 + \alpha_8 \leq 4. \tag{3.15f}$$

The alternative construction needs $\lceil log_2 5 \rceil = 3$ binary variables (Fig. 3.2 highlights region $\mathscr{R}_8^-$ with the corresponding combination of binary variables for both formulations (3.15) or (3.16)):

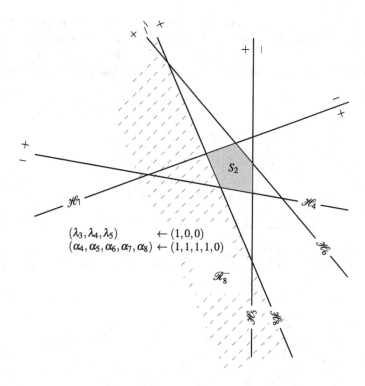

Fig. 3.2 Illustration of mixed-integer formulation for (3.2) in both classical and logarithmic formulation

$$-h_4 x \le -k_4 + M(\lambda_3 + \lambda_4 + \lambda_5), \tag{3.16a}$$

$$-h_5 x \le -k_5 + M(1 + \lambda_3 + \lambda_4 - \lambda_5), \tag{3.16b}$$

$$-h_6 x \le -k_6 + M(1 + \lambda_3 - \lambda_4 + \lambda_5), \tag{3.16c}$$

$$-h_7 x \le -k_7 + M(2 + \lambda_3 - \lambda_4 - \lambda_5), \tag{3.16d}$$

$$-h_8 x \le -k_8 + M(1 - \lambda_3 + \lambda_4 + \lambda_5), \tag{3.16e}$$

$$2 - \lambda_3 + \lambda_4 - \lambda_5 > 0, \tag{3.16f}$$

$$2 - \lambda_3 - \lambda_4 + \lambda_5 > 0, \tag{3.16g}$$

$$3 - \lambda_3 - \lambda_4 - \lambda_5 > 0. \tag{3.16h}$$

Constraints (3.16f)–(3.16h), computed as in Proposition 3.1 make infeasible each of the three remaining unallocated tuples $(1, 0, 1)$, $(1, 1, 0)$ and $(1, 1, 1)$. Graphically, these constraints are illustrated in Fig. 3.3a.

Applying Proposition 3.2 the number of constraints can be further reduced. It can be observed that $(1, 1, 0)$ and $(1, 1, 1)$ share a common value $\lambda_3 = \lambda_4 = 1$ which means that they completely cover a ridge in the $\{0, 1\}^3$ unit hypercube and can be separated by a single constraint. Therefore, the total number of constraints is reduced to 2:

$$2 - \lambda_3 + \lambda_4 - \lambda_5 > 0, \tag{3.17a}$$

$$2 - \lambda_3 - \lambda_4 > 0. \tag{3.17b}$$

Graphically, these constraints are illustrated in Fig. 3.3b.

Lastly, we note that the allocated tuples codify the numbers from 0 to 4 and the unallocated tuples the numbers from 5 to 7. Therefore, as shown in Proposition 3.3 a single constraint suffices for separation:

$$4\lambda_3 + 2\lambda_4 + \lambda_5 \le 4. \tag{3.18}$$

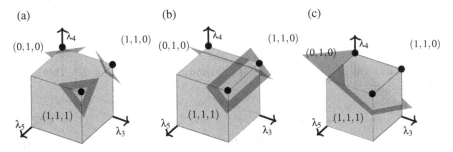

Fig. 3.3 Exemplification of separating hyperplanes techniques

Note that any of the allocated tuples satisfy this constraint and none of the unallocated ones do. Graphically, these constraints are illustrated in Fig. 3.3c.

3.3   The Complement of a Union of Convex Sets

Under construction (3.3a)–(3.3b) the i-th binary variable, α_i, decides whether the i-th inequality is considered (or not) in the optimization problem. In other words, (3.3a)–(3.3b) end up describing the complement of a given polyhedral set. The same can be done for a union of obstacles by simply applying to each element of the union the same construction. In particular, taking (2.21a) as the forbidden region, its complement, region (2.21b) can be defined as:

$$\overline{\mathbb{S}} = \overline{\bigcup_l S_l} = \bigcap_l \overline{S_l} = \bigcap_{\sigma_l^\bullet \in \Sigma^\bullet} \bigcap_{i \in \mathbb{I}} \overline{\mathcal{R}_i^{\sigma_l(i)}} = \bigcap_{\sigma_l^\bullet \in \Sigma^\bullet} \left(\bigcup_{i \in \mathbb{I}} \overline{\mathcal{R}_i^{\sigma_l^\bullet(i)}} \right) = \bigcap_{\sigma_l^\bullet \in \Sigma^\bullet} \left(\bigcup_{i \in \mathbb{I}} \overline{\mathcal{R}_i^{\sigma_l^\bullet(i)}} \right).$$

$$(3.19)$$

This reformulation of the admissible region leads to the following mixed-integer construction.

Proposition 3.4   *For a collection of regions (2.17) characterized by sign tuples $\sigma_l^\bullet \in \Sigma^\bullet$ as in (2.21a) the admissible region (2.21b) can be described by the following mixed-integer formalism as in (3.3a)–(3.3b):*

$$h_i x \leq k_i + M \prod_{\overline{\sigma_l^\bullet(i)} = '+', \forall \sigma_l^\bullet \in \Sigma^\bullet} \alpha_i^{l,\bullet}, \ \forall i \in \mathbb{I}, \tag{3.20a}$$

$$-h_i x \leq -k_i + M \prod_{\overline{\sigma_l^\bullet(i)} = '-', \forall \sigma_l^\bullet \in \Sigma^\bullet} \alpha_i^{l,\bullet}, \ \forall i \in \mathbb{I}, \tag{3.20b}$$

$$\sum_{i \in \mathbb{I}} \alpha_i^{l,\bullet} \leq N - 1, \ \forall \sigma_l^\bullet \in \Sigma^\bullet. \tag{3.20c}$$

□

Proof   Let us recall construction (3.3a)–(3.3b) and apply if for a $\sigma_l^\bullet \in \Sigma^\bullet$. This leads to formulation:

$$\overline{\sigma_l^\bullet(i)} h_i x \leq \overline{\sigma_l^\bullet(i)} k_i + M \alpha_i^{l,\bullet}, \ \forall i \in \mathbb{I}, \tag{3.21a}$$

$$\sum_{i \in \mathbb{I}} \alpha_i^{l,\bullet} \leq N - 1, \tag{3.21b}$$

which signifies that at least a complementary region, $\overline{\mathcal{R}_i^{\sigma_l^\bullet(i)}}$, has to be active for any feasible combination of binary variables $\alpha^{l,\bullet} \in \{0, 1\}^N$. Furthermore, let us assume that inequality $\{h_i x \le k_i\}$ is one of the half-spaces defining $\mathscr{A}(\sigma_1^\bullet)$, $\mathscr{A}(\sigma_2^\bullet)$. Then, using construction (3.21), implies that $\{-h_i x \le -k_i + M\alpha_i^{1,\bullet}\}$ and $\{-h_i x \le -k_i + M\alpha_i^{2,\bullet}\}$, which is equivalent with $\{-h_i x \le -k_i + M\alpha_i^{1,\bullet}\alpha_i^{2,\bullet}\}$. Repeating the same assertions for all $\sigma_l^\bullet \in \Sigma^\bullet$ we reach construction (3.20a)–(3.20c), thus concluding the proof.   ∎

Note that Proposition 3.4 is not the most compact representation as it requires an excessive number of binary variables (N for each of the convex sets considered). In fact, in many situations not all the half-spaces (2.1) are active in a given polyhedral set S_l. We can then reduce the number of binary variables by considering in the description of convex sets only the indices corresponding to non-redundant inequalities:

$$S_l = \bigcap_{i \in \mathbb{I}_l} \mathscr{R}_i^{\sigma_l^\bullet(i)}, \tag{3.22}$$

where $\mathbb{I}_l \subseteq \mathbb{I}$ denotes the collection of indices characterizing the active regions and has $N_l \le N$ elements. This leads to a reformulation of (3.19) as:

$$\overline{\mathbb{S}} = \bigcap_{\sigma_l^\bullet \in \Sigma^\bullet} \left(\bigcup_{i \in \mathbb{I}_l} \mathscr{R}_i^{\overline{\sigma_l^\bullet(i)}} \right), \tag{3.23}$$

which in turn leads to the next corollary of Proposition 3.4.

Corollary 3.2 *For a collection of regions (2.17) characterized by sign tuples as in (2.21a) the admissible region (3.23) can be described by the following mixed-integer formalism as in (3.3a)–(3.3b):*

$$h_i x \le k_i + M \prod_{\overline{\sigma_l^\bullet(i)}='+' \text{ and } i \in \mathbb{I}_l, \forall \sigma_l^\bullet \in \Sigma^\bullet} \alpha_i^{l,\bullet}, \ \forall i \in \mathbb{I}, \tag{3.24a}$$

$$-h_i x \le -k_i + M \prod_{\overline{\sigma_l^\bullet(i)}='-' \text{ and } i \in \mathbb{I}_l, \forall \sigma_l^\bullet \in \Sigma^\bullet} \alpha_i^{l,\bullet}, \ \forall i \in \mathbb{I}, \tag{3.24b}$$

$$\sum_{i \in \mathbb{I}_l} \alpha_i^{l,\bullet} \le N_l - 1, \ \forall \sigma_l^\bullet \in \Sigma^\bullet. \tag{3.24c}$$

Proof The proof is similar to the one of Proposition 3.4 with the difference that the number of binary variables for a region S_l is linked to N_l, the number of active regions characterizing it.   ∎

Some remarks are in order.

Remark 3.6 Note that in both Proposition 3.4 and Corollary 3.2 we do not need the entire description of the hyperplane arrangement (2.4). We only need to enumerate the sign tuples which characterize the forbidden regions (usually fewer than the total number of feasible sign tuples). In fact, we can avoid altogether the hyperplane arrangement construction and simply force that at least one constraint is violated in each of the polyhedral sets S_l.   ◆

Remark 3.7 Having a product of binary variables as in (3.20a)–(3.20b) and (3.24a)–(3.24b) makes the problem nonlinear which is not desirable. In practice, instead of having one inequality with a nonlinear binary term we choose to decompose such that only one term of the product appears in each element. More precisely, inequality $\{hx \leq k + M \prod_j \alpha_j\}$ is in fact equivalent with inequalities $\{hx \leq k + M\alpha_j, \forall j\}$. ◆

Remark 3.8 To each of the constructions from Proposition 3.4 and Corollary 3.2 we can apply the logarithmic construction from Proposition 3.1 with the result of a reduction in the number of binary variables.   ◆

Illustrative example for mixed-integer formulations for the complement of a union of convex sets

Let us continue developing on the illustrative example presented in Chap. 2 (see Fig. 2.5 and Appendix A.4) where the feasible region is represented by the complement of a union of convex sets as depicted in Fig. 3.4. The forbidden region (2.17) is composed of 2 polytopes (the first defined by half-spaces $\{h_i x \leq k_i\}$ with $i = 1, 2, 3$ and the latter by half-spaces $\{h_i x \leq k_i\}$ with $i = 4, 5, 6, 7$).

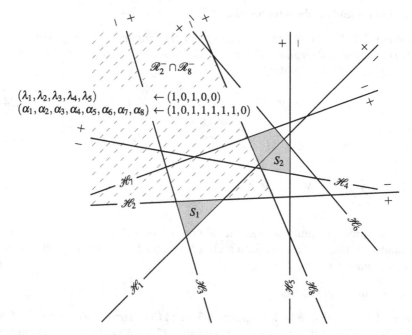

Fig. 3.4 Illustration of mixed-integer formulation for (2.21b) and hyperplane arrangement description

The feasible region (2.21b) is described in the mixed-integer formulation (3.24a)–(3.24c) for each of the polytopes. The result is represented by two groups of inequalities, (3.13) and (3.15) respectively. Enumerating the combinations of binary variables any part of the feasible region can be covered. For example, by taking $(\alpha_1, \alpha_2, \alpha_3) = (1, 0, 1)$ and $(\alpha_4, \alpha_5, \alpha_6, \alpha_7, \alpha_8) = (1, 1, 1, 1, 0)$ we are able to describe $\mathscr{R}_2^- \cap \mathscr{R}_8^-$, as depicted in Fig. 3.4.

The same construction can be repeated for the logarithmic formulation. That is, using (3.14) and (3.16) we are able to again describe the feasible region. In particular, $\mathscr{R}_2^- \cap \mathscr{R}_8^-$, shown in Fig. 3.4 is characterized by $(\lambda_1, \lambda_2) = (1, 0)$ and $(\lambda_3, \lambda_4, \lambda_5) = (1, 0, 0)$.

3.4 Description of the Feasible Region as Union of Feasible Cells

Under construction (3.3a)–(3.3b) the i-th binary variable, α_i, decides whether the i-th inequality is considered (or not) in the optimization problem. The same reasoning holds to the more general case of a group of constraints which is governed by the same binary variable.

If until now we described the feasible region in terms of the complement of the forbidden region we now proceed to a more direct characterization. That is, we directly describe the feasible region as a union of either disjoint cells as in (2.21b) or merged cells as in (2.23).

Recalling (2.21b) we can express the feasible region as:

$$\overline{\mathbb{S}} = \bigcup_{\sigma^\circ \in \Sigma^\circ} \mathscr{A}(\sigma^\circ) = \bigcup_{\sigma^\circ \in \Sigma^\circ} \left(\bigcap_{i \in \mathbb{I}} \mathscr{R}_i^{\sigma^\circ(i)} \right). \tag{3.25}$$

This reformulation of the admissible region leads to the following mixed-integer construction.

Proposition 3.5 *For an admissible region described as in (3.25) we associate the following mixed-integer description:*

$$h_i x \leq k_i + M \prod_{\sigma_l(i) = `+' \, and \, \sigma_l \in \Sigma^\circ} \alpha_l, \, \forall i \in \mathbb{I}, \tag{3.26a}$$

$$-h_i x \leq -k_i + M \prod_{\sigma_l(i) = `-' \, and \, \sigma_l \in \Sigma^\circ} \alpha_l, \, \forall i \in \mathbb{I}, \tag{3.26b}$$

$$\sum_{l=1...N_r} \alpha_l \leq N_r - 1, \tag{3.26c}$$

where $N_r = \text{card}(\Sigma^\circ)$ and denotes the number of sign tuples from (2.20).

Proof Let us recall construction (3.3a)–(3.3b) and extend it for any of the convex sets which are part of (3.25). This leads straightforwardly to the formulation

$$\sigma_l(i)h_i x \leq \sigma_l(i)k_i + M\alpha_l, \ \forall i \in \mathbb{I}, \ \forall \sigma_l \in \Sigma^\circ \tag{3.27a}$$

$$\sum_{l=1...N_r} \alpha_l \leq N_r - 1 \tag{3.27b}$$

where we have that at least an admissible cell $\mathscr{A}(\sigma_l)$ is active for any feasible combination of binary variables. Using the fact that two inequalities which have the same real part can be concatenated we then reach construction (3.26a)–(3.26c), thus concluding the proof.   ∎

Having obtained the merged cells of (2.22) it becomes straightforward to modify Proposition 3.5 as per the next corollary using the following construction:

$$\overline{\mathbb{S}} = \bigcup_{\sigma^* \in \Sigma^*} \mathscr{A}(\sigma^*) = \bigcup_{\sigma^* \in \Sigma^*} \left(\bigcap_{i \in \mathbb{I}, \sigma^*(i) \neq '*'} \mathscr{R}_i^{\sigma^*(i)} \right). \tag{3.28}$$

Corollary 3.3 *For an admissible region described as in (3.28) the following mixed-integer description is associated:*

$$h_i x \leq k_i + M \prod_{\sigma^*(i)='+' \text{ and } \sigma^* \in \Sigma^\circ} \alpha_l, \ \forall i \in \mathbb{I}, \tag{3.29a}$$

$$-h_i x \leq -k_i + M \prod_{\sigma^*(i)='-' \text{ and } \sigma^* \in \Sigma^\circ} \alpha_l, \ \forall i \in \mathbb{I}, \tag{3.29b}$$

$$\sum_{l=1...N_r^*} \alpha_l \leq N_r^* - 1, \tag{3.29c}$$

where N_r^* denotes the number of sign tuples characterizing merged cells.

Proof The proof is similar to the one of Proposition 3.5 with the difference that the number of binary variables is equal to the number of merged cells, and that some of the hyperplanes are not active.   ∎

This leads to a more compact representation (and as the most important consequence, to less binary variables) than in (3.20a)–(3.20c) with obvious benefits for the computation times. The number of binary variables can be further reduced by employing the logarithmic formulation of Proposition 3.1.

Illustrative example for mixed-integer formulations for the union of feasible cells

Let us recall the previous illustrative example (see the numerical data provided in Appendix A.4). Also, recall that in the illustrative example of Sect. 2.3.1 we gave the list of sign tuples which correspond to the merged cells (2.22), computed either via Theorem 2.2 or Corollary 2.1.

In what follows a mixed-integer representation of (3.25) is constructed with the tools presented above.

Using the list of disjoint cells characterizing the hyperplane arrangement (see Chap. 2) and according to Proposition 3.5 the following mixed-integer representation is obtained:

$$
\left.
\begin{array}{l}
-h_1 x \leq -k_1, \\
h_2 x \leq k_2, \\
h_3 x \leq k_3, \\
-h_4 x \leq -k_4, \\
h_5 x \leq k_5, \\
h_6 x \leq k_6, \\
h_7 x \leq k_7, \\
-h_8 x \leq -k_8,
\end{array}
\right\} + M\alpha_1,
\tag{3.30a}
$$

$$
\dots \text{equations for the remaining 33 cells,} \tag{3.30b}
$$

$$
\sum_{i=1}^{34} \alpha_i \leq 33. \tag{3.30c}
$$

For compactness reasons, Fig. 3.5 delineates only one cell, $\mathscr{A}(-++-++ +-)$. In (3.30) the term $M\alpha(1)$ corresponding to the illustrated cell controls whether the cell is or is not active. The same construction is repeated for the remaining 33 feasible cells describing $\overline{\mathbb{S}}$. Lastly, condition (3.30c) ensures that at least one of the cells remains active.

Next, the merged cell decomposition (3.28) is employed by using the list of merged cells obtained in the illustrative example of Sect. 2.3.1: $\mathscr{A}(*-*****-)$, $\mathscr{A}(-**-****)$, $\mathscr{A}(*****-**)$, $\mathscr{A}(**-*****)$, $\mathscr{A}(******-*)$ and $\mathscr{A}(****-***)$. For these cells a mixed-integer description is obtained as shown in Corollary 3.3. Here, for a more compact representation, a logarithmic formulation is proposed:

$$
\left.
\begin{array}{l}
-h_2 x \leq -k_2 \\
-h_8 x \leq -k_8
\end{array}
\right\} + M(\lambda_1 + \lambda_2 + \lambda_3),
\tag{3.31a}
$$

$$
\left.
\begin{array}{l}
-h_1 x \leq -k_1 \\
-h_4 x \leq -k_4
\end{array}
\right\} + M(1 + \lambda_1 + \lambda_2 - \lambda_3),
\tag{3.31b}
$$

$$
-h_6 x \leq -k_6\} + M(1 + \lambda_1 - \lambda_2 + \lambda_3), \tag{3.31c}
$$

$$
-h_3 x \leq -k_3\} + M(2 + \lambda_1 - \lambda_2 - \lambda_3), \tag{3.31d}
$$

$$
-h_7 x \leq -k_7\} + M(1 - \lambda_1 + \lambda_2 + \lambda_3), \tag{3.31e}
$$

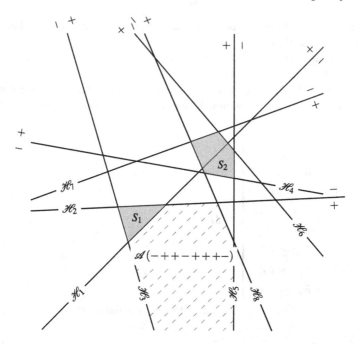

Fig. 3.5 Illustration of mixed-integer formulation as in Proposition 3.5 and hyperplane arrangement description

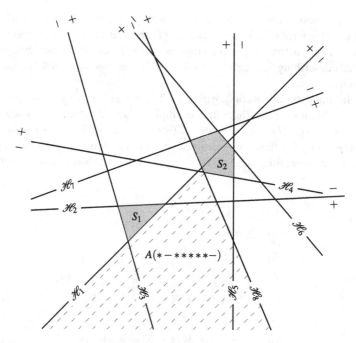

Fig. 3.6 Illustration of mixed-integer formulation as in Corollary 3.3 and hyperplane arrangement description

$$-h_5 x \leq -k_5\} + M(2 - \lambda_1 + \lambda_2 - \lambda_3), \tag{3.31f}$$

$$4\lambda_1 + 2\lambda_2 + \lambda_3 \leq 5.5. \tag{3.31g}$$

Figure 3.6 delineates the merged cell $\mathscr{A}(* - * * * * *-)$ described by (3.31a).

Note that in both cases the description of the merged cells can be done using both the classical and the logarithmic construction. For the disjoint cell construction the 34 feasible cells can be codified through 6 binary variables ($2^6 = 64 > 33$) whereas for the merged cell construction 3 binary variables suffice ($2^3 = 8 > 6$). The auxiliary constraint in (3.31g) makes the unallocated sign tuples (i.e., $(1, 1, 0)$ and $(1, 1, 1)$) infeasible as explained in Sect. 3.2.2.

3.5   Description of the Feasible Region Directly Through the Arrangement

Up to this point the hyperplane arrangements have been used in an auxiliary role, i.e., to generate codifications for the regions of interest (either the ones forbidden or the ones admitted). While we maintain that using hyperplane arrangement notions is a superior way of obtaining polyhedral decompositions, it is also true that any other decomposition method which describes these regions would do. The next approach makes use of the arrangement structure and is embedded into the mixed-integer representation.

The justification for the third way of representing the feasible region is to define it as the "region which is not inside the obstacles":

$$\bar{\mathbb{S}} = \mathbb{R}^n \setminus \bigcup_l S_l = \mathbb{R}^n \setminus \bigcup_{\sigma_l^\bullet \in \Sigma^\bullet} \mathscr{A}(\sigma_l^\bullet). \tag{3.32}$$

This reformulation of the admissible region leads to the following MI construction.

Proposition 3.6 *Consider the forbidden region (2.17) and its associated hyperplane arrangement (2.4). Then, the feasible region (3.32) is given by the following mixed-integer formulation:*

$$h_i x \leq k_i + M(1 - \alpha_i), \tag{3.33a}$$

$$-h_i x \leq -k_i + M\alpha_i, \tag{3.33b}$$

$$\sum_{\sigma_l^\bullet(i)='+'} (1 - \alpha_i) + \sum_{\sigma_l^\bullet(i)='-'} \alpha_i > 0, \ \forall \sigma_l^\bullet \in \Sigma^\bullet. \tag{3.33c}$$

Proof Constraints (3.33a)–(3.33b) ensure that either region $\mathscr{R}_i^+$ or $\mathscr{R}_i^-$ is active, depending on the value taken by the associated binary variable α_i. Considering only (3.33a)–(3.33b) means that the entire space will be covered, including the forbidden regions (2.17). To avoid this, we add conditions (3.33c) which, by making infeasible the sign tuples associated with forbidden cells, ensure that only the cells describing (3.32) are feasible.   ∎

Remark 3.9 In general, the mixed-integer representation (3.20a)–(3.20c) should be more compact than (3.33a)–(3.33c) in the number of necessary binary variables (especially if for the former we use the logarithmic formulation). Nonetheless, the latter's construction is interesting since the number of binary variables remains fix, no matter how many cells are labeled as forbidden. Note that the number of interdiction constraints (3.33c) can be further reduced if by a single constraint more than one point is made infeasible (see Proposition 3.2 of [3]).   ◆

Illustrative example for mixed-integer formulations for the feasible region characterized directly through the arrangement

Continuing the set of illustrative examples using the arrangement given in Appendix A.4 we recall that three cells are forbidden, i.e., $\Sigma^\bullet = \{(+++-++ +-), (-+++ + ++), (-++++ --+)\}$. Therefore, Proposition 3.6 provides the following mixed-integer formulation:

$$h_i x \leq k_i + M(1 - \alpha_i), \; \forall i = 1 \ldots 8, \tag{3.34a}$$
$$-h_i x \leq -k_i + M\alpha_i, \; \forall i = 1 \ldots 8, \tag{3.34b}$$
$$\alpha_1 + \alpha_2 + \alpha_3 + (1 - \alpha_4) + \alpha_5 + \alpha_6 + \alpha_7 + (1 - \alpha_8) > 0, \tag{3.34c}$$
$$(1 - \alpha_1) + \alpha_2 + \alpha_3 + \alpha_4 + \alpha_5 + \alpha_6 + \alpha_7 + \alpha_8 > 0, \tag{3.34d}$$
$$(1 - \alpha_1) + \alpha_2 + \alpha_3 + \alpha_4 + \alpha_5 + (1 - \alpha_6) + (1 - \alpha_7) + \alpha_8 > 0. \tag{3.34e}$$

3.6 Numerical Considerations

This section tests the constructive methods enumerated earlier using the numerical data from Appendices A.3–A.9. In particular, the following constructions which characterize the feasible domain are considered:

Ia – as a complement of the obstacles, as in Proposition 3.4;
Ib – as a complement of the obstacles, as in Corollary 3.2;
Ic – as a complement of the obstacles, as in Proposition 3.4 with logarithmic representation;
Id – as a complement of the obstacles, as in Corollary 3.2 with logarithmic representation;
IIa – as a union of feasible disjoint cells, as in Proposition 3.5 with logarithmic representation;
IIb – as a union of feasible merged cells, as in Corollary 3.3;
IIc – as a union of feasible merged cells, as in Corollary 3.3 with logarithmic representation;
III – via the hyperplane arrangement, as in Proposition 3.6.

Note the use of the shorthand notation of the methods: I, II and III stand for the big design choices shown in Sects. 3.1–3.3 and the small letters for variation within the same class of methods (whether the classical or the logarithmic formulation is employed or if redundant half-spaces are taken out of the problem characterization).

There are two main issues of practical interest in the implementation of these mixed-integer constructions. The first is related to the memory footprint and the second to the computation time.

To illustrate the problem size, the arrangements from Appendices A.1–A.9 are considered and each of them are put into the mixed-integer constructions enumerated above. For each of these instants the number of constraints and of binary variables is counted. The first number shows the size of the problem whereas the second is a proxy for the problem difficulty.

A short remark regarding the number of binary variables in a mixed-integer problem is in order. A dedicated solver will travel inside a decision tree where each branch represents a different combination of binary values. Hence, a large number of binary variables can lead (at least theoretically) to an exponential computation time. In practice, modern solvers can tackle efficiently the problem. Still, the number of binary variables in a mixed-integer problem remains as an useful indicator of worst-case complexity.

Table 3.1 shows the number of constraints and binary variables corresponding to each arrangement under the various mixed-integer constructions enumerated above (each cell has two numbers, the first is the number of constraints and the second denotes the number of binary variables).

As expected, the cases Id, IIc and III are the most compact (due to the elimination of redundant information, use of the logarithmic formulation, construction particularities). Note as well that case IIa becomes impractical to represent at a relatively small problem dimension.

In Table 3.2 the count of the constraints and binary variables is repeated but for a larger number of obstacles ($\Sigma^{\bullet,2}$ defined in Appendix A).

Again, the most compact representation are found in cases Id, IIc and III. For these particular examples it seems that the number of binary variables increases in

Table 3.1 Number of constraints and binary variables for various hyperplane arrangements

(N, d)	(3,2)	(8,2)	(10,2)	(15,2)	(20,2)	(25,2)	(30,2)
Case Ia	(4,3)	(27,24)	(44,40)	(144,135)	(189,180)	(806,775)	(682,660)
Case Ib	(4,3)	(15,12)	(18,14)	(48,39)	(48,39)	(154,123)	(116,94)
Case Ic	(4,2)	(27,9)	(44,16)	(144,36)	(189,45)	(806,155)	(682,110)
Case Id	(4,2)	(15,7)	(18,8)	(48,22)	(48,21)	(154,69)	(116,50)
Case IIa	(114,3)	(4632,5)	(17682,6)	(118904,7)	*	*	*
Case IIb	(12,3)	(54,6)	(112,8)	(198,11)	(180,10)	(817,19)	(512,16)
Case IIc	(10,2)	(48,3)	(104,3)	(184,4)	(168,4)	(778,5)	(484,4)
Case III	(7,3)	(19,8)	(24,10)	(39,15)	(49,20)	(81,25)	(82,30)

Table 3.2 Number of constraints and binary variables for various hyperplane arrangements (with a larger number of obstacles)

(N, d)	(3,2)	(8,2)	(10,2)	(15,2)	(20,2)	(25,2)	(30,2)
Case Ia	(8,6)	(45,40)	(77,70)	(368,345)	(294,280)	(1040,1000)	(1147,1110)
Case Ib	(7,5)	(24,19)	(34,27)	(114,91)	(74,60)	(200,160)	(194,157)
Case Ic	(8,4)	(45,15)	(77,28)	(368,92)	(294,70)	(1040,200)	(1147,185)
Case Id	(7,3)	(24,11)	(34,16)	(114,51)	(74,34)	(200,90)	(194,84)
Case IIa	(80,3)	(3894,5)	(15249,6)	(84450,7)	*	*	*
Case IIb	(6,2)	(84,7)	(98,7)	(240,10)	(242,11)	(1020,20)	(1134,21)
Case IIc	(5,1)	(79,3)	(89,3)	(222,4)	(224,4)	(975,5)	(1085,5)
Case III	(8,3)	(21,8)	(27,10)	(53,15)	(54,20)	(90,25)	(97,30)

small increments (with the exception of case III where the number of binary variables remains, by construction, constant).

In the rest of this section we test the actual computation times. We proceed as follows: for a given arrangement, we take 100 random points inside the union of obstacles and write a QP optimization problem whose goal is to minimize the distance between the solution and the current infeasible point subject to the constraints of the mixed-integer problem. The constraints come from the mixed-integer constructions of cases Id, IIa, IIb, IIc and III. We average the computation time and illustrate it in the Tables 3.3 and 3.4.

As in the previous chapter, in order to have a better grasp of the computation times for the mixed-integer procedures under the various formulations, Fig. 3.7 plots the relevant results provided in Table 3.4.

Several remarks are in order. First of all, the computation times remain small with the exception of case IIa where the excessive number of constraints leads to an exponential increase in the computation time. Otherwise, note that the union of disjoint cells with logarithmic representation is equivalent or faster than the merged case (case IIa against cases IIb and IIc). The logarithmic formulation is faster but with a relatively small improvement w.r.t. the classic case. Lastly, for these examples it seems that exploiting the arrangement yields the smallest computation times. With respect to the variation between 'less obstacles' and 'many obstacles' the computation

Table 3.3 Computations times for various hyperplane arrangements

(N, d)	(3,2)	(8,2)	(10,2)	(15,2)	(20,2)	(25,2)	(30,2)
Case Id	0.23751	0.23213	0.2358	0.23758	0.23392	0.26213	0.2502
Case IIa	0.25646	0.6331	1.482	5.7714	*	*	*
Case IIb	0.23475	0.25868	0.28026	0.31392	0.29395	0.47498	0.40342
Case IIc	0.23275	0.2561	0.27749	0.30611	0.28796	0.4571	0.39211
Case III	0.22887	0.23299	0.23438	0.23254	0.22701	0.24111	0.23543

Table 3.4 Computations times for various hyperplane arrangements (with a larger number of obstacles)

(N, d)	(3,2)	(8,2)	(10,2)	(15,2)	(20,2)	(25,2)	(30,2)
Case Id	0.23258	0.23152	0.22993	0.25011	0.2406	0.27478	0.27117
Case IIa	0.24346	0.55676	1.2546	4.0993	*	*	*
Case IIb	0.22731	0.26027	0.25812	0.29759	0.31122	0.50991	0.53791
Case IIc	0.22618	0.25858	0.25479	0.29115	0.30379	0.48961	0.51592
Case III	0.22575	0.22688	0.22601	0.23726	0.23247	0.24914	0.24684

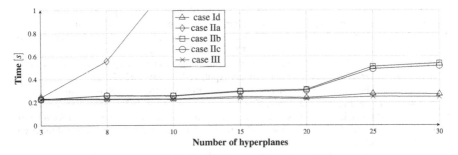

Fig. 3.7 Illustration of the mixed-integer representations computation times provided in Table 3.4

time vary in small increments. That is, for any of the methods, the times are higher but not with large amounts.

To summarize, all three main methods (in brief, 'complement', 'union', 'arrangement') provide comparable results in terms of computation times and seem to be "fast enough" for reasonable size problems.

Lastly, a word of caution is in order: the results will vary significantly with the solver used. In these examples the IBM ILOG CPLEX 12.6.1 solver has been used. Other solvers might be less adept at pruning the decision tree/discarding redundant inequalities. Therefore, not one of these constructive methods should be considered "the best". Rather, the user should test with his/her own solver, observe the behavior and choose accordingly.

3.7 Notes and Comments

This chapter addresses the problem of minimizing the computational complexity of optimization problems under a non-convex and possibly non-connected feasible region. Using the hyperplane arrangements notions we provide a compact description of the feasible region and thus obtain a compact mixed-integer formulation with fewer binary variables. We emphasize the various MIP constructions possibilities and discuss their relative advantages and disadvantages. These constructions range from these which merely use the hyperplane arrangements to provide a description of the feasible region and up to these which use the intrinsic structure of the arrangement to define the mixed-integer problem.

Some of the above contributions have been briefly presented in previous papers [3, 5, 6]. Somewhat related topics have been discussed in [4, 7].

References

1. Jünger, M., Junger, M., Liebling, T., Naddef, D., Nemhauser, G., Pulleyblank, W.: 50 Years of Integer Programming 1958-2008: From the Early Years to the State-of-the-Art. Springer (2009)
2. Vielma, J., Nemhauser, G.: Modeling disjunctive constraints with a logarithmic number of binary variables and constraints. Math. Program. 128(1), 49–72 (2011)
3. Prodan, I., Stoican, F., Olaru, S., Niculescu, S.I.: Enhancements on the hyperplanes arrangements in mixed-integer techniques. J. Optim. Theory Appl. 154(2), 549–572 (2012)
4. Afonso, R.J., Galvão, R.K.: Comments on "Enhancements on the hyperplanes arrangements in mixed-integer programming techniques". J. Optim. Theory Appl. 162(3), 996–1003 (2014)
5. Stoican, F., Prodan, I., Olaru, S.: On the hyperplanes arrangements in mixed-integer techniques. In: Proceedings of the 30th American Control Conference, pp. 1898–1903. San Francisco, California, USA (2011)
6. Stoican, F., Prodan, I., Olaru, S.: Enhancements on the hyperplane arrangements in mixed integer techniques. In: Proceedings of the 50th IEEE Conference on Decision and Control and European Control Conference, pp. 3986–3991. Orlando, Florida, USA (2011)
7. Geyer, T., Torrisi, F., Morari, M.: Optimal complexity reduction of polyhedral piecewise affine systems. Automatica 44(7), 1728–1740 (2008)

Chapter 4
Control Problems Involving Mixed-Integer Decision Making

This chapter validates the results presented in the previous chapters over two well-known and still largely open classes of control problems. Namely, we apply the mixed-integer formalism to the *multi-agent control* and *fault tolerant control* topics. These are particularly interesting as many of the typical goals reduce to constrained optimization problems over non-convex regions.

For example, for the multi-agent class of control problems there is usually a collaborative cost function characterizing some desired common goal (formation flight, reaching of a destination and the like) while in the same time obstacles and other agents have to be avoided. Alternatively, we consider topics which are specific to multi-agent dynamical systems but have not really been tackled from the control perspective. In particular, we are interested by motion planning with *obstacle and collision avoidance*, *coverage* and *corner cutting* problems. All these topics (and others not discussed here) can be written in the mixed-integer formalism and moreover, can be improved through the use of the hyperplane arrangement decomposition discussed earlier.

A multi-agent system, at least in the control setting, has a lax definition. It is defined by any group of agents (robots, vehicles or pedestrians) which participate in a physical environment and whose aim is to accomplish cooperatively common design objectives [1].

There are various methods in the literature for solving multi-agent collision and obstacles avoidance problems. Arguably, they can be gathered in methods which penalize through the cost function the violation of the constraints (e.g., Potential Field Method [2] and Navigation Functions [3]) and methods which impose hard constraints that may not be broken. The latter group usually employs receding horizon techniques as they naturally take into account constraints [4–7].

In the present chapter, the evolution of a dynamical agent in an environment presenting obstacles is modeled in terms of a non-convex feasible region. More precisely, we set up an optimization problem such that the agent state trajectory avoids

© The Author(s) 2016
I. Prodan et al., *Mixed-Integer Representations in Control Design*,
SpringerBriefs in Control, Automation and Robotics,
DOI 10.1007/978-3-319-26995-5_4

some convex region, in fact, representing an obstacle (static constraints) or another agent (dynamic constraints leading to a parametrization of the set of constraints with respect to the current state).

The multi-agent coverage and corner cutting problems reduce to the use of *shadow regions* generated by the agents and obstacles and they can be defined in general as a position assignment problem responding to the question: *where should the agents be positioned in the feasible space such that all the space is observed?*

For the above mentioned problems, the mixed-integer formulation is then integrated in a *constrained optimization-based control formulation* which is a MPC (Model Predictive Control) (see, for example [8–10]) design.

For the second class of applications, the FTC problem, the main purpose is to automatically attenuate/cancel the negative effects of a component fault. Any FTC scheme relies on two fundamental mechanisms: the FDI (Fault Detection and Isolation) block and the RC (control reconfiguration) block. The solutions employed usually implement *active* FTC schemes which react to a detected fault and reconfigure the control actions so that stability and performance can be satisfied.

Usually [11], the detection and reconfiguration parts of a FTC scheme are treated separately thus neglecting reciprocal influences and substandard behavior (e.g., missed faults). The proposed scheme, based on invariant set separation [12, 13] and further detailed in [14], integrates all the FTC components, and analyzes their interactions, to create an overall system with guaranteed fault tolerance properties.

Any FDI mechanism is, at its most general, a threshold validation mechanism which checks a specially designed signal (the residual) for unexpected behavior. This assumes sufficient excitation in the closed-loop system, such that in case of model mismatch due to fault occurrences, the differences between expected and observed behavior are significant. Such an excitation is usually provided by a reference which is assumed to lie in a predefined bound. This is constrictive as the reference is treated as an external signal with no relation to the fault tolerant control desiderata. Thus, we take here the reference as a decision variable which is subject to fault detection and isolation restrictions [14]. This means that the feasible domain (i.e., the region in which the FDI restrictions are valid) is a non-convex region to which the previous mixed-integer constructions can be applied.

4.1 Multi-agent Collision Avoidance Problems

Let us consider a collection of N_a dynamical agents and without significant loss of generality let us assume that they are homogeneous, i.e., they are characterized by the same dynamics:

$$x_i(k+1) = Ax_i(k) + Bu_i(k), \quad y_i(k) = Cx_i(k), \tag{4.1}$$

with $x_i(k) \in \mathbb{R}^n$, $u_i(k) \in \mathbb{R}^m$ and $y_i(k) \in \mathbb{R}^p$ representing the i-th agent state, input and output, respectively.

Furthermore, let us consider a collection of obstacles, $\mathbb{S}$, with the notation from Chap. 3. In this framework, a straightforward task for the multi-agent system is motion planning with collision free behavior. More precisely, each agent has to navigate its way around the obstacles while in the same time avoiding collision with the other agents. Note that in many situations the avoidance constraints are imposed on a subspace of the state space, i.e., on the output space. The most readily available example is autonomous vehicle motion planning where usually avoidance constraints are imposed only on position subspace.

Even if in simulation an agent is usually simplified to a point, in real-life applications it might have a shape which is significant with respect to the obstacle shapes [15]. Also, the state of the agent is usually computed for the nominal dynamics but the presence of disturbances and model mismatches implies that the 'real' state will actually stay near but not quite in the 'nominal' state. All these issues represent strong reasons for associating to each agent a *safety region*. This region can either be given as an external parameter or computed based on noise and perturbations bounds [16, 17]. For further use, the safety region of the i-th agent is denoted by S_i^a.

Having all the necessary ingredients, let us formulate in the following the collision and avoidance conditions:

(i) for any obstacle S_l and any agent characterized by its dynamical state $x_i(k)$ and the associated safety region S_i^a, the collision avoidance conditions are:

$$\left(\{x_i(k)\} \oplus S_i^a\right) \cap S_l = \emptyset, \quad \forall i = 1 \dots N_a, \ \forall l = 1 \dots N_o. \quad (4.2)$$

(ii) for any two agents characterized by their dynamical states $x_i(k), x_j(k)$ and their associated safety regions S_i^a, S_j^a, the collision avoidance conditions are:

$$\left(\{x_i(k)\} \oplus S_i^a\right) \cap \left(\{x_j(k)\} \oplus S_j^a\right) = \emptyset, \quad \forall i, j = 1 \dots N_a, \ i \neq j. \quad (4.3)$$

It is worth mentioning that a zonotope cannot in general approximate arbitrary well all convex shapes (as a polytope does) but can, nonetheless, provide over-approximations of these shapes (recall Sect. 2.2.2). This approach can be considered here for a collection of safety regions associated to the agents and obstacles. These sets are defined as $\mathscr{Z}(x_i(k), G_i^a)$ and $\mathscr{Z}(o_l, G_l)$, respectively. That is, around each agent $x_i(k)$ with $i \in \{1 \dots N_a\}$ there is a zonotopic safety region with generators G_i^a and each obstacle $l \in \{1 \dots N_o\}$ is described by a fix center o_l and the generators G_l.

If the generators' directions are fixed, an optimization problem can scale them such that they over-approximate efficiently a given shape [18]. Using the same construction here we assume a common "seed", the generators G, and Δ_i^a and Δ_l the scaling factors (diagonal matrices with positive elements) for agent i and obstacle l, respectively. Then, using the above zonotopic constructions, the collision and obstacle avoidance conditions as in (4.2)–(4.3) are formulated as follows:

$$\mathscr{Z}(x_i(k), G\Delta_i^a) \cap \mathscr{Z}(o_l, G\Delta_l) = \emptyset, \quad \forall i = 1 \ldots N_a, \; \forall l = 1 \ldots N_o, \tag{4.4a}$$

$$\mathscr{Z}(x_i(k), G\Delta_i^a) \cap \mathscr{Z}(x_j(k), G\Delta_j^a) = \emptyset, \quad \forall i, j = 1 \ldots N_a, \; i \neq j. \tag{4.4b}$$

Using the fact that $(\{a\} \oplus A) \cap (\{b\} \oplus B) = \emptyset$ is equivalent with $\{a - b\} \notin \{-A\} \oplus B$, the avoidance conditions (4.2)–(4.3) are rewritten as:

$$x_i(k) \notin \left(\{-S_i^a\} \oplus S_l\right), \quad \forall i = 1 \ldots N_a, \; \forall l = 1 \ldots N_o, \tag{4.5a}$$

$$x_i(k) - x_j(k) \notin \left(\{-S_i^a\} \oplus S_j^a\right), \quad \forall i, j = 1 \ldots N_a, \; i \neq j. \tag{4.5b}$$

Similarly, using the zonotope properties (2.15)–(2.16) an equivalent formulation for (4.4a)–(4.4b) is reached:

$$x_i(k) \notin \mathscr{Z}(o_l, G(\Delta_i^a + \Delta_l)), \quad \forall i = 1 \ldots N_a, \; \forall l = 1 \ldots N_o, \tag{4.6a}$$

$$x_i(k) - x_j(k) \notin \mathscr{Z}(0, G(\Delta_i^a + \Delta_j^a)), \quad \forall i, j = 1 \ldots N_a, \; i \neq j. \tag{4.6b}$$

Let us now consider the set of N_a constrained systems as a *global system* defined as:

$$\mathbf{x}(k+1) = \mathbf{A}\mathbf{x}(k) + \mathbf{B}\mathbf{u}(k), \quad \mathbf{y}(k) = \mathbf{C}\mathbf{x}(k), \tag{4.7}$$

with the corresponding vectors which collect the states, the inputs and the outputs of each individual agent (4.1) at time k, i.e., $\mathbf{x}(k) \triangleq \left[x_1^\top(k) \ldots x_{N_a}^\top(k)\right]^\top$, $\mathbf{u}(k) \triangleq \left[u_1^\top(k) \ldots u_{N_a}^\top(k)\right]^\top$ and $\mathbf{y}(k) \triangleq \left[y_1^\top(k) \ldots y_{N_a}^\top(k)\right]^\top$, respectively, and the matrices which describe the centralized dynamical model: $\mathbf{A} = diag[\underbrace{A, \ldots, A}_{N_a}]$,

$\mathbf{B} = diag[\underbrace{B, \ldots, B}_{N_a}], \; \mathbf{C} = diag[\underbrace{C, \ldots, C}_{N_a}].$

Furthermore, an optimal control problem is considered for the global system where the cost function and the constraints couple the dynamic behavior of the individual agents. Also, perfect knowledge of each agent dynamics (4.1) is available to all the other agents.

Typically, a MPC approach is preferred as it allows for the explicit consideration of costs, constraints and references. In particular, this framework allows to consider a non-convex feasible domain (as resulting from the collision and avoidance constraints described by (4.5a)–(4.5b) or (4.6a)–(4.6b)).

We consider a cost function $V(\mathbf{x}, \mathbf{u}) : \mathbb{R}^{N_a \cdot n} \times \mathbb{R}^{N_a \cdot m} \rightarrow \mathbb{R}$ which aims at maintaining a formation, following a reference path or simply to gather the agents towards the origin and the constraints defined as in (4.5a)–(4.5b) or (4.6a)–(4.6b). A finite receding horizon implementation is typically based on the solution of an open-loop optimization problem. An optimal control action $\mathbf{u}^*$ is obtained from the control sequence $\mathbf{u} \triangleq \{u(k), u(k+1), \ldots, u(k + N_p - 1)\}$ as a result of the optimization problem:

$$\mathbf{u}^* = \arg \min_{\mathbf{u}(k)...\mathbf{u}(k+N_p-1)} \sum_{t=0}^{N_p-1} V(\mathbf{x}(k+t), \mathbf{u}(k+t)), \tag{4.8a}$$

$$\text{s.t.:} \quad \mathbf{x}(k+t+1) = \mathbf{A}\mathbf{x}(k+t) + \mathbf{B}\mathbf{u}(k+t), \quad t = 0 : N_p - 1, \tag{4.8b}$$

$$\mathbf{y}(k+t) = \mathbf{C}\mathbf{x}(k+t), \quad t = 1 \ldots N_p, \tag{4.8c}$$

$$\mathbf{x}(k+t) \notin \mathbb{S}, \quad t = 1 \ldots N_p. \tag{4.8d}$$

The optimization problem (4.8) requires the minimization of the cost function over a finite prediction horizon N_p. Condition (4.8d) comes from (4.5a)–(4.5b) or (4.6a)–(4.6b) which can be readily recognized as leading to a non-convex feasible domain. These constraints reduce to "$\mathbf{x}(k)$ outside a union of convex sets", denoted by $\mathbb{S}$. From the optimal sequence of inputs $\mathbf{u}(k)^*, \ldots, \mathbf{u}(k + N_p - 1)^*$ the first control input, $\mathbf{u}(k)^*$, is selected and applied to the global system, thus closing the loop.

Remark 4.1 The problem can be further simplified by noting that the agents' dynamics are usually subject to operational constraints (e.g., magnitude or rate of variation constraints) which limit their actual range. This means that we need to consider in the description of $\overline{\mathbb{S}}$ only the cells which are intersecting the *reachable set* of the agents. The simplification of the scheme can be significant, e.g., if no obstacle is in the "line of sight" of the agents and the agents themselves are far away from each other, then the resulting feasible domain will be convex.   ♦

Lastly, note that the resulting constrained optimization problem describes a centralized scheme, decentralized or distributed schemes are possible (see, for example [15]) but are beyond the scope of this monograph.

Illustrative example for collision avoidance constraints

Let us consider here the LTI dynamics of the "unicycle model" (commonly used to describe simplified vehicle dynamics through their position and velocity—[19]):

$$x(k+1) = \underbrace{\begin{bmatrix} I_2 & I_2 \\ 0_2 & I_2 \end{bmatrix}}_{A} x(k) + \underbrace{\begin{bmatrix} 0_2 \\ I_2 \end{bmatrix}}_{B} u(k), \quad y(k) = \underbrace{\begin{bmatrix} I_2 & 0_2 \end{bmatrix}}_{C} x(k), \tag{4.9}$$

which is a discretized integrator (with sampling time $\Delta t = 1$) with state composed from position and velocity and the output given as the agent's position. I_2 and 0_2 are the identity and zero matrices from $\mathbb{R}^2$, respectively.

The multi-obstacle environment defined by the hyperplane arrangement from Appendix A.4 is considered here. For illustration purposes we take a single agent which has no safety region, has to avoid the obstacles and reach the origin. The MPC problem has a quadratic cost $\sum_{t=0}^{N_p-1} \left(||\mathbf{x}(k+t)||_Q + ||\mathbf{u}(k+t)||_R \right)$ with the weighting matrices $Q = \text{diag}(I_2, 0_2)$ and $R = I_2$, the prediction horizon $N_p = 10$ and the output, and respectively, the input constraints of form $\mathscr{Y} = \{y : |y| \leq \begin{bmatrix} 15 & 15 \end{bmatrix}^\top\}$ and $\mathscr{U} = \{u : |u| \leq \begin{bmatrix} 0.5 & 0.5 \end{bmatrix}^\top\}$.

Table 4.1 Computation times for various starting positions under several feasible space decompositions for the arrangement from Appendix A.4

Start	1	2	3	4	5	6	7	8	9	10
Case Id	1.76	2.07	1.95	1.48	1.68	1.30	1.74	2.13	2.35	2.59
Case IIa	14.31	14.31	14.30	13.40	14.61	16.52	15.84	26.65	18.24	17.49
Case IIb	1.22	1.29	1.18	1.28	1.25	1.36	1.24	1.74	1.46	1.38
Case IIc	1.45	1.48	1.55	1.50	1.43	1.51	1.49	1.85	1.79	1.59
Case III	2.59	3.07	2.71	2.73	2.65	3.68	4.59	3.19	4.51	3.60

We consider 10 random starting positions which lie outside the obstacles (we assume the velocity is initialized at '0') and we propagate the trajectory starting from each of them with the input provided by the constrained optimization problem (4.8) for 15 simulation steps. The 10 initial points are $(-10.15, -7.45)$, $(-10.02, 2.41)$, $(-2.76, 0.73)$, $(-6.38, -11.97)$, $(10.36, -11.46)$, $(12.96, -1.77)$, $(11.59, 3.28)$, $(12.44, 9.71)$, $(7.60, 9.93)$, $(2.25, 11.24)$.

For each of these runs we have considered multiple decompositions of the feasible space (with notation from Sect. 3.6, cases Id, IIa, IIb, IIc and III) and depicted in Table 4.1 the total computation times for each of these methods.

Case IIa, which takes a union of disjoint cells to characterize the space, is by far the slowest approach. The remaining ones are relatively close with the fastest being cases IIb and IIc which require a pre-processing merging procedure. On the other hand cases Id and III, which do not require a significant pre-processing step, are relatively close in terms of performance. Also of note is that within each method there are significant variations with respect to the starting position of the trajectory run.

For illustration, Fig. 4.1 delineates the starting points and the resulting trajectories of the agent. Note that each of the space decompositions results (usually) in identical trajectories and therefore, we show here only the state trajectories resulted from the application of case III. The only differences appear to be near the obstacle boundary in some of the runs of case Id.

Of further interest is how these computation times modify with increased number of obstacle within the same hyperplane arrangement and increased number of hyperplanes. The first situation is illustrated in Table 4.2 and the second in Table 4.3.

Table 4.2 shows that there are increases in the computation times but not significant, at least for the relatively simple example of Appendix A.4. On the other hand, in Table 4.3 we observe a marked increase in the computation times. Case IIa has proved impossible to solve but the other cases exhibit as well a performance degradation.

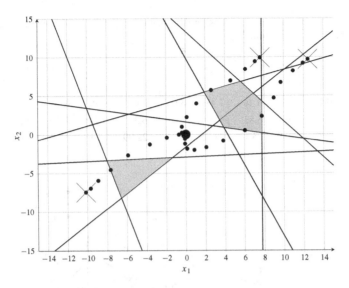

Fig. 4.1 Illustration of obstacle avoidance for an agent

Table 4.2 Computation times for various starting positions under several feasible space decompositions for the arrangement from Appendix A.4, with a larger number of obstacles

Start	1	2	3	4	5	6	7	8	9	10
Case Id	2.15	1.76	1.93	2.47	2.48	4.36	2.76	3.84	3.14	4.32
Case IIa	7.86	4.24	3.96	6.85	10.08	18.82	13.24	25.62	15.21	8.87
Case IIb	1.35	1.31	1.73	1.49	1.59	1.63	1.59	1.82	1.48	1.56
Case IIc	1.63	1.51	1.60	1.77	2.05	1.91	1.92	1.92	1.85	1.87
Case III	2.53	2.94	4.54	3.29	2.88	3.26	3.99	3.45	3.98	3.66

Figure 4.2 illustrates the obstacle and collision avoidance of a multi-agent system. Three agents, each with its own safety region (defined as a unit square box centered in the agent's state, $x_i(k)$) are subjected to the same constraints and cost as for the earlier single agent case.

It can be observed that along the simulation horizon both collision and obstacle avoidance conditions for the multi-agent system are respected (i.e., the safety regions do not intersect with each other or with the obstacles). Furthermore, at the end of the simulation, a tight formation around the origin is reached. This comes from the presence of the safety regions and deserves further attention [16, 20].

Note that the formulation can be further simplified along the lines (4.6a)–(4.6b) by over-approximating the obstacles through zonotopic sets. For further details see [21].

Table 4.3 Computation times for various starting positions under several feasible space decompositions for the arrangement from Appendix A.9

Start	1	2	3	4	5	6	7	8	9	10
Case Id	17.39	67.66	15.33	11.71	11.03	13.30	25.47	18.46	27.82	26.96
Case IIa	*	*	*	*	*	*	*	*	*	*
Case IIb	5.00	5.75	6.34	7.48	6.06	5.01	6.49	6.46	5.94	5.89
Case IIc	6.53	6.20	6.83	6.02	6.05	6.30	6.28	6.25	6.20	6.11
Case III	21.42	25.59	25.19	23.85	20.38	22.99	20.18	21.75	24.18	24.77

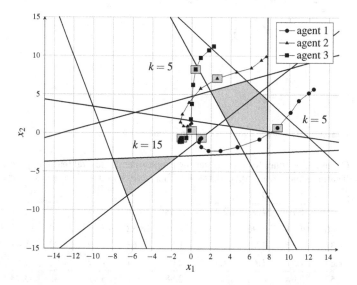

Fig. 4.2 Illustration of obstacle and collision avoidance for multiple agents

4.2 Extensions for Multi-agent Motion Planning

In the context of multi-agent control systems within a multi-obstacle environment several apparently disparate topics share an underlying framework. Line-of-sight communication, coverage of the feasible space, corner cutting, all of these require a parametrization of the visible/under shadow regions generated by one or more agents against a collection of obstacles.

We will demonstrate that these regions are parametrized after the hyperplane arrangement associated to the obstacles. Besides a better understanding of the geometry of the problem this also allows for better problem formulations. Further, we may consider over-approximations which, at the price of being more conservative, allow for simpler coding of the constraints. We note that the hyperplane arrangement is not only a convenient way to describe the obstacles but also serves as support for the shadow area function (i.e., the structure of the shadow region remains constant over a fixed cell) which henceforth allows a piecewise description.

To bring the previous constructions into a manageable form mixed-integer constructions are employed. This allows to both simplify the formulation (see Chap. 2) with respect to the existing results in the literature and to exploit the piecewise nature of the shadow regions [22]. Together with codification methods found in Chap. 3 and the references therein we provide mixed-integer constructs which describe the shadow region and its complement in both exact and over-approximation forms. For the latter, a simplified form involving only binary variables is achieved.

4.2.1 Shadow Region Description

Assuming the obstacle notation from the previous chapters let us consider a point $x \in \mathbb{R}^n \backslash \mathbb{S}$. Then, the *shadow region* $\mathcal{B}(\mathbb{S}, x)$ given as in [22] is the collection of all the points from $\mathbb{R}^n \backslash \mathbb{S}$ which are not "visible" from 'x':

$$\mathcal{B}(\mathbb{S}, x) = \{ y \in \mathbb{R}^n : \quad [x, y] \cap \mathbb{S} \neq \emptyset \}. \tag{4.10}$$

In other words, if the segment $[x, y]$ intersects $\mathbb{S}$ it means that point x is "hidden" by obstacles $\mathbb{S}$ and therefore is not "visible" from the viewpoint of x.

Considering definition (2.21a), region (4.10) is rewritten[1] as:

$$\mathcal{B}(\mathbb{S}, x) = \bigcup_{\sigma^\bullet \in \Sigma^\bullet} \mathcal{B}(\sigma^\bullet, x). \tag{4.11}$$

To construct the set (4.11) we have to deal with the parameter x. In order to do so, consider hereinafter the auxiliary construction:

$$\mathcal{E}(\sigma^\bullet, x) = \mathcal{A}(\sigma^\bullet) \cap \left(\bigcup_{x \notin \mathcal{H}_i^{\sigma^\bullet(i)}} \mathcal{H}_i \right) \cap \left(\bigcup_{x \in \mathcal{H}_i^{\sigma^\bullet(i)}} \mathcal{H}_i \right), \tag{4.12}$$

which denotes the tangent points of $\mathcal{A}(\sigma^\bullet)$ from the viewpoint of x.

Remark 4.2 In $\mathbb{R}^2$, (4.12) reduces to a collection of disconnected extreme points of the obstacle $\mathcal{A}(\sigma^\bullet)$. In general, in $\mathbb{R}^n$ we obtain a connected union of $n - 1$ flats ("ridges"). ♦

Proposition 4.1 *For any $x \in \mathcal{A}(\sigma^\circ)$ where $\sigma^\circ \in \Sigma^\circ$, the shadow region $\mathcal{B}(\sigma^\bullet, x)$ has a constant structure given by:*

$$\mathcal{B}(\sigma^\bullet, x) = \text{Cone}(x, \mathcal{E}(\sigma^\bullet, x)) \cap \bigcap_{\sigma^\circ(i) \neq \sigma^\bullet(i)} H_i^{\sigma^\bullet(i)}, \tag{4.13}$$

---

[1]To shorten the notation, we write $\mathcal{B}(\mathcal{A}(\sigma^\bullet), x)$ in the compact form $\mathcal{B}(\sigma^\bullet, x)$.

where[2]

$$\mathscr{E}(\sigma^{\bullet}, \sigma^{\circ}) = \mathscr{A}(\sigma^{\bullet}) \cap \left(\bigcup_{\sigma^{\circ}(i) \neq \sigma^{\bullet}(i)} \mathscr{H}_i \right) \cap \left(\bigcup_{\sigma^{\circ}(i) = \sigma^{\bullet}(i)} \mathscr{H}_i \right). \tag{4.14}$$

□

Proof By construction, the shadow area can be written as $\mathscr{B}(\sigma^{\bullet}, x) = \text{Cone}$ $(x, \mathscr{A}(\sigma^{\bullet})) \cap \bigcap_{x \notin H_i^{\sigma^{\bullet}(i)}} H_i^{\sigma^{\bullet}(i)}$. This becomes (4.13) if we note that the cone spanned from x and tangent to $\mathscr{A}(\sigma^{\bullet})$ is completely characterized by x and $\mathscr{E}(\sigma^{\bullet}, x)$. Term (4.12) is rewritten in form (4.14) if we note that the indices for which $x \notin \mathscr{H}_i^{\sigma^{\bullet}(i)}$ and $x \in \mathscr{H}_i^{\sigma^{\bullet}(i)}$ remain the same for any point taken from $\mathscr{A}(\sigma^{\circ})$ and are in fact given by checking whether the regions $\mathscr{A}(\sigma^{\bullet})$ and $\mathscr{A}(\sigma^{\circ})$ lie on the same (or opposite) sides of the i-th hyperplane.   ■

Proposition 4.1 shows that it suffices to compute a parametrized set (4.13) for any x in a given cell $\mathscr{A}(\sigma^{\circ})$ and then replace the parameter x with the actual value at run-time. While this reduces the computation burden, the formulation for the shadow area is still relatively difficult due to Cone $(x, \mathscr{E}(\sigma^{\bullet}, \sigma^{\circ}))$. A solution, as shown in the next corollary, is to consider an over-approximation of the shadow region.

Corollary 4.1 *Let there be* $\mathscr{B}(\sigma^{\bullet}, \sigma^{\circ}) = \bigcup_{x \in \mathscr{A}(\sigma^{\circ})} \mathscr{B}(\sigma^{\bullet}, x)$ *the shadow region associated to a feasible tuple* σ°. *Then, this region depends only on* σ° *and is described as follows:*

$$\mathscr{B}(\sigma^{\bullet}, \sigma^{\circ}) = \bigcap_{\sigma^{\circ}(i) \neq \sigma^{\bullet}(i)} H_i^{\sigma^{\bullet}(i)}, \tag{4.15}$$

□

Proof From the definition of $\mathscr{B}(\sigma^{\bullet}, \sigma^{\circ})$, the fact that $\bigcup_i (A_i \cap B) = \left(\bigcup_i A_i \right) \cap B$ and (4.13) follows that $\mathscr{B}(\sigma^{\bullet}, \sigma^{\circ}) = \bigcup_{x \in \mathscr{A}(\sigma)} \text{Cone}(x, \mathscr{E}(\sigma^{\bullet}, \sigma^{\circ})) \cap \bigcap_{\sigma^{\circ}(i) \neq \sigma^{\bullet}(i)} H_i^{\sigma^{\bullet}(i)}$ which leads to (4.15).   ■

By using the over-approximation (4.15) the shadow region not only retains the same structure for any $x \in \mathscr{A}(\sigma^{\circ})$ but actually remains constant. Hence, at run-time it is necessary only to identify the currently active tuple σ° and use the corresponding region (4.15).

Remark 4.3 In general, we may consider the shadow area resulting from a set rather than from a point ($x \in \mathbf{X}$). The only difficulty is to check whether the set $\mathbf{X}$ stays in one or more of the regions (2.21b). Defining $\Sigma_{\mathbf{X}} \triangleq \{\sigma \in \Sigma^{\circ} : \text{interior}(\mathbf{X} \cap \mathscr{A}(\sigma)) \neq \emptyset\} \subseteq \Sigma^{\circ}$ allows to characterize the shadow region:

---

[2]To underline that the set depends only on σ° and not on any particular $x \in \mathscr{A}(\sigma^{\circ})$ we changed from $\mathscr{E}(\sigma^{\bullet}, x)$ to $\mathscr{E}(\sigma^{\bullet}, \sigma^{\circ})$.

$$\mathscr{B}(\sigma^{\bullet}, \mathbf{X}) = \bigcup_{x \in \mathbf{X} \cap \mathscr{A}(\sigma_x), \sigma_x \in \Sigma_{\mathbf{X}}} (\mathscr{B}(\sigma^{\bullet}, x)),$$ (4.16a)

$$\mathscr{B}(\sigma^{\bullet}, \Sigma_{\mathbf{X}}) = \bigcup_{\sigma_{\mathbf{X}} \in \Sigma_{\mathbf{X}}} \mathscr{B}(\sigma^{\bullet}, \sigma_{\mathbf{X}}).$$ (4.16b)

along the lines of Proposition 4.1 and Corollary 4.1. ◆

Illustrative example for shadow regions

Consider the numerical example presented in Chap. 2 where a collection of obstacles in $\mathbb{R}^2$, $\mathbb{S} = S_1 \cup S_2$ and their associated hyperplane arrangement was presented (see also Fig. 2.5 with the numerical data provided in Appendix A.4). These partition the space into 34 cells from which 3 describe the 2 obstacles and the rest characterize the feasible space $\mathbb{R}^2 \backslash \mathbb{S}$. More precisely, $\Sigma^{\bullet} = \{\sigma_1, \sigma_2, \sigma_3\}$ is identified such that $S_1 = \mathscr{A}(\sigma_1)$ and $S_2 = \mathscr{A}(\sigma_2) \cup \mathscr{A}(\sigma_3)$ for $\sigma_1 = (+ + + - + + + -)$, $\sigma_2 = (+ - + + + + + +)$ and $\sigma_3 = (- - + + + + + +)$.

We apply now the shadow region descriptions for S_1. In Fig. 4.3 we take a point $x_1 \in \mathscr{A}(\sigma)$ for $\sigma = (- + + - + + + -)$.

Checking the signs for tuples σ_1 and σ we note that x_1 shares the same half-spaces with S_1 for all indices bar index '1'. Using this information in (4.12), or alternatively in (4.14), we note that the set of tangent points of S_1 from the viewpoint

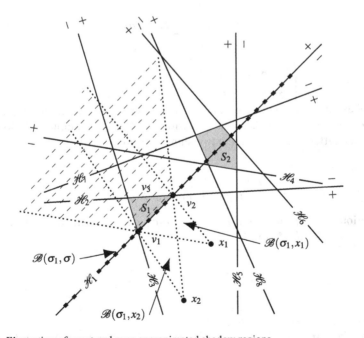

Fig. 4.3 Illustration of exact and over-approximated shadow regions

of x_1 consists of two points: $\mathcal{E}(\sigma_1, x_1) = \{\mathcal{H}_1 \cap \mathcal{H}_1, \mathcal{H}_1 \cap \mathcal{H}_3\}$. Next, we can compute Cone(x_1, S_1), which is the cone defined by the rays starting from x_1 and passing through $\mathcal{E}(\sigma_1, x_1)$. By adding the half-spaces separating the obstacle from the observation point (in this case, there is only one separating half-space, $\mathcal{R}_1^+$) we obtain the shadow region $\mathcal{B}(\sigma_1, x_1) = \text{Cone}(x_1, S_1) \cap \mathcal{R}_1^+$. We depict in Fig. 4.3 the over-approximation strategy employed in Corollary 4.1. That is, we take an additional point $x_2 \in \mathcal{A}(\sigma)$ from the same cell and depict the resulting shadow region $\mathcal{B}(\sigma_1, x_2) = \text{Cone}(x_2, S_1) \cap \mathcal{R}_1^+$. Comparing with $\mathcal{B}(\sigma_1, \sigma) = \mathcal{R}_1^+$, constructed as in Corollary 4.1, it can be seen that $\mathcal{B}(\sigma_1, \sigma)$ contains any region $\mathcal{B}(S_1, x)$ for $x \in \mathcal{A}(\sigma)$, and in particular for $x \in \{x_1, x_2\}$.

4.2.2 Mixed-Integer Representations

Earlier we gave various formulations for shadow regions observed from the point of view of an agent and with multiple obstacles. Regardless of the particular construction, the issue is that the resulting feasible region is non-convex (and in the case of multiple obstacles, not even connected). Henceforth, we use the mixed integer formulations from Chap. 3 which help describe the problem in a pseudo-linear formulation.

As a first step, we define the mapping $\delta : \{-, +\}^N \times \{-, +\}^N \to \{0\} \cup [1, \infty)$:

$$\delta(\sigma_1, \sigma_2) = \sum_i \delta_i(\sigma_1, \sigma_2), \tag{4.17}$$

where $\delta_i(\sigma_1, \sigma_2) = \begin{cases} 1 - \sigma_2(i), & \sigma_1(i) = \text{`}+\text{'} \\ \sigma_2(i), & \sigma_1(i) = \text{`}-\text{'} \end{cases}$.

We abuse the notation and whenever convenient (in the definition of (4.17) for example) we equate '−' with '0' and '+' with '1'. With this in mind, if follows that δ maps[3] the difference between the two sign tuples σ_1 and σ_2:

$$\delta(\sigma_1, \sigma_2) = \begin{cases} 0, & \sigma_1 = \sigma_2 \\ \geq 1, & \sigma_1 \neq \sigma_2 \end{cases}.$$

We give now the following representation of $\mathcal{B}(\sigma^\bullet, x)$.

Proposition 4.2 *Let there be a point $x \in \mathcal{A}(\sigma)$ and an obstacle $\mathcal{A}(\sigma^\bullet)$. Then a point x^+ is inside $\mathcal{B}(\sigma^\bullet, x)$ iff*

---

[3]This can be intuitively read as $\delta(\sigma_1, \sigma_2) = \sum_i |\sigma_1(i) - \sigma_2(i)|$. The mappings $\delta(\cdot, \cdot)$ and $\delta_i(\cdot, \cdot)$ are preferred because they highlight the linearity of the description.

$$x^+ = x + \sum_j \beta_j (v_j - x), \tag{4.18a}$$

$$\sum_{j \text{ s.t. } v_j \notin \mathscr{E}(\sigma^\bullet, \sigma^\circ)} \beta_j \leq M\delta(\sigma^\circ, \sigma), \quad \forall \sigma^\circ \in \Sigma^\circ, \tag{4.18b}$$

$$\beta_j \geq 0, \tag{4.18c}$$

$$\sigma^\bullet(i)h_i x^+ \leq \sigma^\bullet(i)k_i + M(1 - \delta_i(\sigma^\bullet, \sigma)), \tag{4.18d}$$

where v_j denote the extreme points of the obstacle. □

Proof The detailed proof is to be found in [23]. ■

Remark 4.4 Proposition 4.2 assumes that the current location of point x (i.e., its sign tuple σ) is unknown. This is the reason for which we enumerate all the feasible tuples in (4.18b). Subsequently, iff $\sigma \neq \sigma^\circ$ the associated inequality is neglected (due to the 'big-M' formulation). ♦

In Proposition 4.2 we provided a mixed-integer description of the shadow area. However, if we wish to describe the feasible region, we need its complement.

Proposition 4.3 *Let there be a point $x \in \mathscr{A}(\sigma)$ and an obstacle $\mathscr{A}(\sigma^\bullet)$. Then a point x^+ is outside $\mathscr{B}(\sigma^\bullet, x)$ iff*

$$M(1 - \alpha) \geq |x^+ - x - \sum_j \beta_j(v_j - x)|, \tag{4.19a}$$

$$|\beta_j| \leq M\delta(\sigma^\circ, \sigma), \forall j \text{ s.t. } v_j \notin \mathscr{E}(\sigma^\bullet, \sigma^\circ), \tag{4.19b}$$

$$\beta_j \leq M(1 - \gamma_j), \tag{4.19c}$$

$$\gamma_j \leq \delta(\sigma^\circ, \sigma), \forall j \text{ s.t. } v_j \notin \mathscr{E}(\sigma^\bullet, \sigma^\circ), \tag{4.19d}$$

$$\sum_j \gamma_j > 0, \tag{4.19e}$$

$$-\sigma^\bullet(i)h_i x^+ \leq -\sigma^\bullet(i)k_i + M[\delta_i(\sigma^\bullet, \sigma) + \alpha + \rho_i], \tag{4.19f}$$

$$N > \sum_i \left[\rho_i + \delta_i(\sigma^\bullet, \sigma)\right], \tag{4.19g}$$

for any $\sigma^\circ \in \Sigma^\circ$ and with $\alpha, \gamma_j, \rho_i \in \{0, 1\}$ auxiliary binary variables. □

Proof The detailed proof is to be found in [23]. ■

Remark 4.5 Note that in both Proposition 4.2 and 4.3 we considered only one obstacle (the one defined by tuple $\sigma^\bullet$). The extension to the case of multiple obstacles is straightforward. For the shadow area additional binary variables need to be considered in order to describe the union of shadow regions resulting from each of the obstacles. On the other hand, to describe the visible region, we simply intersect the regions obtained in (4.19a)–(4.19g): $x \notin \bigcup_{\sigma^\bullet \in \Sigma^\bullet} B(\sigma^\bullet, x)$ is equivalent with

$$x \in \overline{\bigcup_{\sigma^\bullet \in \Sigma^\bullet} B(\sigma^\bullet, x)} = x \in \bigcap_{\sigma^\bullet \in \Sigma^\bullet} \overline{B(\sigma^\bullet, x)}.$$ ◆

The exact formulations (4.18a)–(4.18d) and (4.19a)–(4.19g) lead to complex representations. This is due to the presence of term $\text{Cone}(\sigma^\bullet, x)$. If on the other hand we use the over-approximation of Corollary 4.1, we greatly simplify the representations.

Proposition 4.4 *Assume that the current position is $x \in \mathscr{A}(\sigma)$ and that the obstacle is $\mathscr{A}(\sigma^\bullet)$. Then the future position $x^+ \in \mathscr{A}(\sigma^+)$ is constrained as follows:*

(i) for $x^+ \in \mathscr{B}(\sigma^\bullet, \sigma)$:

$$\sum_i \delta_i(\sigma^\bullet, \sigma) \cdot \delta_i(\sigma^\bullet, \sigma^+) = 0, \tag{4.20}$$

(ii) for $x^+ \notin \mathscr{B}(\sigma^\bullet, \sigma)$:

$$\sum_i \delta_i(\sigma^\bullet, \sigma) \cdot \delta_i(\sigma^\bullet, \sigma^+) > 0. \tag{4.21}$$

□

Proof We can constrain the region of interest by enumerating the half-spaces which separate between the obstacle and the current position. Assuming that the region of interest is described by sign tuple σ^+ it follows that it is constrained by the current position (described by σ) and the obstacle (described by $\sigma^\bullet$).

Whenever $\sigma^\bullet(i)$ and $\sigma(i)$ share the same sign, $\delta_i(\sigma^\bullet, \sigma) = 0$ and if they have opposite signs, $\delta_i(\sigma^\bullet, \sigma) = 1$. The same holds for $\sigma^\bullet(i)$ and $\sigma^+(i)$ w.r.t. $\delta_i(\sigma^\bullet, \sigma^+)$. Using these elements we may characterize the region of interest.

Depending which region we wish to describe (shadow or visible region) we reach (4.20) or (4.21). For case (i), forcing the equality means that each of the terms of the sum is zero (since we have a sum of positive terms) and hence σ^+ describes the shadow region. To describe case (ii) it suffices that at least one of the terms of the sum is non-zero. This is done through (4.21). Again, because each of the sum terms is positive, the fact that the sum is positive means that at least one of them is positive.

The detailed proof is to be found in [23]. ∎

In Proposition 4.4 the constraints (4.20)–(4.21) are linear only if σ is known. If σ is itself a variable we have bilinear binary terms which make any optimization problem into which they appear MINLP. Needless to say, this should be avoided at all costs. The solution is to provide a piecewise description which, through an increase in the number of constraints, keeps the formulation linear.

Corollary 4.2 *Assume that the obstacle is $\mathscr{A}(\sigma^\bullet)$. Then the future position $x^+ \in \mathscr{A}(\sigma^+)$ is constrained as follows:*

(i) for $x^+ \in \mathscr{B}(\sigma^\bullet, \sigma)$:

$$\sum_i \delta_i(\sigma^\bullet, \sigma^\circ) \cdot \delta_i(\sigma^\bullet, \sigma^+) \leq N\delta(\sigma^\circ, \sigma), \tag{4.22}$$

(ii) *for* $x^+ \notin \mathcal{B}(\sigma^\bullet, \sigma)$:

$$\sum_i \delta_i(\sigma^\bullet, \sigma^\circ) \cdot \delta_i(\sigma^\bullet, \sigma^+) > -N\delta(\sigma^\circ, \sigma), \tag{4.23}$$

for all $\sigma^\circ \in \Sigma^\circ$. □

Proof We write the equations (4.20)–(4.21) for each of the feasible sign tuples $\sigma^\circ \in \Sigma^\circ$. Further, recall that $\delta(\sigma^\circ, \sigma)$ is ≥ 1 iff $\sigma^\circ \neq \sigma$. It follows that out of the equations (4.22)–(4.23) only the ones corresponding to the active sign tuple remain and the rest are ignored (as they are always true, regardless of the left-hand side value). ∎

Remark 4.6 Note that (4.22)–(4.23) are not 'big-M' formulations. We are certain that the left-hand sides of the equations cannot be larger than 'N'—the number of hyperplanes and hence we do not need to relax the right-hand term to a value larger than that. ◆

Illustrative example for mixed-integer formulations for the shadow and visible regions

We consider the example from Sect. 4.2.1 and recall that the shadow region spanned from point x_1 with respect to obstacle $S_1 = \mathcal{A}(\sigma_1)$ was found to be:

$$\mathcal{B}(\sigma_1, x_1) = \text{Cone}(x_1, S_1) \cap \mathcal{H}_1^+.$$

Using Proposition 4.2 we obtain the following mixed integer formulation (as seen in Fig. 4.4, the extreme points of S_1 are denoted as v_1, v_2, v_3):

$$x^+ = x + \beta_1(v_1 - x) + \beta_2(v_2 - x) + \beta_3(v_3 - x),$$

$$\cdots\cdots$$

$$\beta_3 \leq M\,(3 - \sigma(1) + \sigma(2) + \sigma(3) - \sigma(4) + \sigma(5) + \sigma(6) + \sigma(7) - \sigma(8)),$$

$$\cdots\cdots$$

$$\beta_1 \geq 0, \ \beta_2 \geq 0, \ \beta_3 \geq 0,$$

$$-h_i x^+ \leq -k_i + M(1 - \sigma(i)), \ i \in \{4, 8\},$$

$$h_i x^+ \leq k_i + M\sigma(i), \ i \in \{1, 2, 3, 5, 6, 7\}.$$

It can be observed that the cone can be composed from at most three rays (the extreme points of S_1). Which of these is active is governed by Eq. (4.18b). For compactness reasons we show here only the equations corresponding to $\sigma^\circ = (- + + - + + + -)$. The last eight constraints show which of the half-spaces of the obstacle remain active in the shadow region representation, constructed as in (4.18d). Consider that in the above equations σ is replaced with σ°, then the following effects are observed:

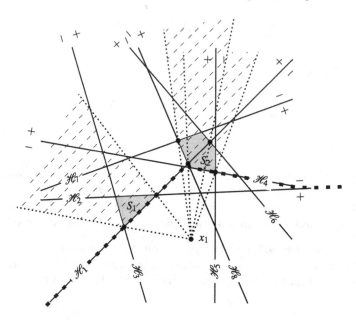

Fig. 4.4 Illustration of over-approximated shadow regions in a multi-obstacle environment

(1) the right side of the third equation becomes zero and hence $\beta_3 = 0$ which means that the cone formulation becomes $x^+ = x + \beta_1(v_1 - x) + \beta_2(v_2 - x)$; (2) out of the half-space constraints we remain with only $h_1 x^+ \leq k_1$. We can then conclude that whenever $x_1 \in \mathcal{A}(\sigma^\circ)$ the shadow region is $\mathcal{B}(\sigma_1, x_1) = \mathrm{Cone}(x_1, S_1) \cap \mathcal{R}_1^+$.

Proposition 4.3 has a similar construction as the one provided in the previous illustrative example, hence we skip to the over-approximate representations given in Proposition 4.4. We consider $\mathcal{B}(\sigma^\bullet, \sigma^\circ)$ with the sign tuples defined as in (4.15). Then, the shadow region inclusion constraint is given as $1 - \sigma^+(1) = 0$ and the shadow region exclusion constraint as $1 - \sigma^+(1) > 0$.

This can be easily checked in Fig. 4.4. From the first equations we have that x^+ can lie in any region which respects $\sigma^+(1) = $ '+', in other words $x^+ \in \mathcal{R}_1^+$ whereas the second equation imposes the opposite, that x^+ cannot lie in any region which respects $\sigma^+(1) = $ '+', hence that $x^+ \notin \mathcal{R}_1^+$. These constraints can easily manipulate multiple obstacles. The collection of constraints:

$$\begin{cases} 0 &= 1 - \sigma^+(1), \\ &\mathrm{OR} \\ 0 &= (1 - \sigma^+(1)) + \sigma^+(2) + (1 - \sigma^+(4)) + (1 - \sigma^+(8)), \\ &\mathrm{OR} \\ 0 &= \sigma^+(2) + (1 - \sigma^+(4)) + (1 - \sigma^+(8)), \end{cases}$$

describes $\mathscr{B}(\mathbb{S}, \sigma) = \mathscr{R}_1^+ \cup \left(\mathscr{R}_1^+ \cap \mathscr{R}_2^- \cap \mathscr{R}_4^+ \cap \mathscr{R}_8^+ \right) \cup \left(\mathscr{R}_2^- \cap \mathscr{R}_4^+ \cap \mathscr{R}_8^+ \right)$ the overall shadow region generated by $x_1 \in \mathscr{A}(\sigma)$ w.r.t. $\mathbb{S}$. Three inequalities appear because S_2 is characterized by two forbidden tuples.

The collection of constraints:

$$
\begin{cases}
0 & < 1 - \sigma^+(1), \\
& \text{AND} \\
0 & < (1 - \sigma^+(1)) + \sigma^+(2) + (1 - \sigma^+(4)) + (1 - \sigma^+(8)), \\
& \text{AND} \\
0 & < \sigma^+(2) + (1 - \sigma^+(4)) + (1 - \sigma^+(8)),
\end{cases}
$$

describes $\mathbb{R}^2 \backslash \mathscr{B}(\mathbb{S}, \sigma) = \mathscr{R}_1^- \cap \left(\mathscr{R}_1^- \cup \mathscr{R}_2^+ \cup \mathscr{R}_4^- \cup \mathscr{R}_8^- \right) \cap \left(\mathscr{R}_2^+ \cup \mathscr{R}_4^- \cup \mathscr{R}_8^- \right)$, the overall visible region generated by $x_1 \in \mathscr{A}(\sigma)$ w.r.t. $\mathbb{S}$.

The thick dotted contour shown in Fig. 4.4 separates the overall shadow region (to the left) and the overall visible region (to the right). The extension to Corollary 4.2 is straightforward and thus not exemplified here.

4.2.3 Coverage Problem

Multi-agent area coverage represents an active research topic due to its wide applicability in various scenarios such as military operations for search and rescue [24], astronomical observations and explorations [25], mobile ad-hoc sensor networks [26] or in general environment monitoring [27]. Likewise, significant interest was shown in the mathematical community on giving bounds for the minimal number of agents [28] and on various topological conditions (e.g., the "art gallery problem" [29]).

In here we make again use of the constructions from Sect. 4.2.1. Considering a collection of agents, described by their position x_k (or by their sign tuples σ_k) we observe that each of them "sees" a region $\mathbb{R}^n \backslash \mathscr{B}(\sigma^\bullet, x_k)$ (or $\mathbb{R}^n \backslash \mathscr{B}(\sigma^\bullet, \sigma_k)$ in the over-approximate form). Put together, these regions define the *overall visible region*

$$
\bigcup_k \left[\bigcap_{\sigma^\bullet \in \Sigma^\bullet} \left(\mathbb{R}^n \backslash \mathscr{B}(\sigma^\bullet, \sigma_k) \right) \right] = \mathbb{R}^n \backslash \left[\bigcap_k \left(\bigcup_{\sigma^\bullet \in \Sigma^\bullet} \left(\mathscr{B}(\sigma^\bullet, \sigma_k) \right) \right) \right], \tag{4.24}
$$

where each point of this region is observed by at least one of the agents. Ideally this region should coincide with the feasible space ($\mathbb{R}^n \backslash \mathbb{S}$), or as much as possible of it, if not.

For the over-approximate case (4.24), the fact that a given collection of sign tuples $\{\sigma_k\}$ completely covers the space reduces to an infeasibility condition. If there does not exist a σ s.t.

$$\mathscr{A}(\sigma) \in \bigcap_k \left[\bigcup_{\sigma^\bullet \in \Sigma^\bullet} (\mathscr{B}(\sigma^\bullet, \sigma_k)) \right], \tag{4.25a}$$

$$\delta(\sigma^\bullet, \sigma) > 0, \forall \sigma^\bullet \in \Sigma^\bullet, \tag{4.25b}$$

then the collection $\{\sigma_k\}$ completely covers the space. (4.25a) limits sign tuple 'σ' to stay in the un-covered regions (the ones which are not observed by any of the agents). In addition (4.25b) does not let σ be any of the sign tuples describing the obstacles. Hence, if the problem is still feasible it means that the shadow area contains part(s) of the feasible space. Conversely, if the problem is infeasible then there is no part of the feasible domain $\mathbb{R}^n \setminus \mathbb{S}$ which is "under shadow".

Variables σ_k can be interpreted in either a dynamic or static key. For the former they represent static positions for a collection of agents which simultaneously cover the feasible space. Otherwise, they can be seen as way-points along the path taken by an agent.

Remark 4.7 For both problems discussed in this section, the existence of a solution is an open issue. In the first case, it may be that the number N_a of agents is insufficient for a complete coverage. On the other hand, in the second problem, the path passing through the way-points also has to respect the agent's dynamic and state/input constraints.   ◆

Remark 4.8 These two cases represent extreme situations of the coverage problem. Various in-between cases can be also considered, e.g., a group of agents cannot cover the entire space simultaneously but has to do it in a minimal time/with a minimal cost of operation.   ◆

Illustrative example for the coverage problem

Taking the previous example we obtain a collection of three sign tuples which guarantee complete coverage: $\sigma_1 = (-++-+++-)$, $\sigma_2 = (+--++++--)$ and $\sigma_3 = (+-++----+)$. The region covered by each of these is depicted in Fig. 4.5.

Putting together all these regions we observe that no cell of the feasible space remains uncovered. As a static problem this requires three agents (note that their positions inside the cells are arbitrary and can be the subject of further optimization). As a dynamic problem, it is an allocation and path finding problem: find the shortest/best path which passes through the pre-defined way-points.

4.2.4 Corner Cutting Problem

Avoidance constraints are usually imposed at the sampling time without regards to the intra-sample behavior of the agent. This means that it is possible for an agent to

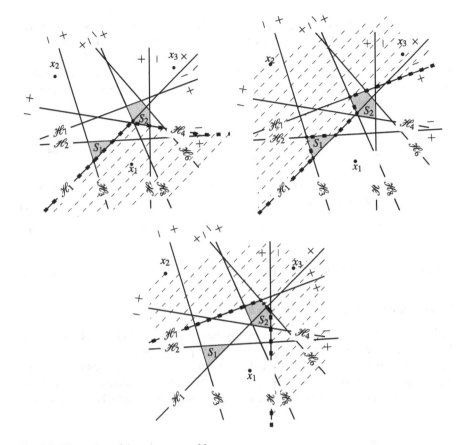

Fig. 4.5 Illustration of the coverage problem

"cut the corner" of an obstacle while apparently respecting the constraints. This type of behavior is usually handled by changing the constraints to take into account the future position of the agent with respect to its current position. There are a variety of ways of implementing these requirements but they all seem (to the best of the authors' knowledge) to be conservative in their description [30–32].

Within our framework this can be seen as the requirement that the next state of the agent will not lie inside the shadow regions generated by the current state of the agent.

Assuming, as it reasonable to do for small sampling times, that in a sampling interval an agent moves along a straight line we can avoid cutting the corner of an obstacle by forcing the next position of the agent to be outside of the shadow region spanned by the current position and the obstacles. In an MPC context, as detailed in Sect. 4.1, we have to check this condition along the entire prediction horizon. That is, we reach a constraint of form $y(k+i+1) \notin \bigcup_{\sigma^\bullet \in \Sigma^\bullet} \mathscr{B}(\sigma^\bullet, y(k+i))$—or its over-approximation counterpart $\mathscr{B}(\sigma^*, \sigma_k)$—in addition to the usual constraints. The sign tuple σ_{k+i} characterizing the predicted output $y(k + i)$ is itself a variable and cannot be assumed as a constant. Here we make use of the previous sections where we have provided piecewise descriptions of the shadow regions (either through Proposition 4.3 or Corollary 4.2). With these constructs we can now formulate the optimization problem and solve it:

$$u^* = \arg \min_{u(k),\sigma_{k+1},\ldots u(k+N_p-1),\sigma_{k+N_p}} \sum_{i=0}^{N_p-1} \|x(k + i + 1)\|_Q + \|u(k)\|_R, \qquad (4.26a)$$

$$\text{s.t.} x(k + i + 1) = Ax(k + i) + Bu(k + i), \qquad (4.26b)$$

$$y(k + i) \in \mathscr{Y}, \ u(k + i) \in \mathscr{U}, \qquad (4.26c)$$

$$y(k + i + 1) \notin \bigcup_{\sigma^* \in \Sigma^*} \mathscr{B}(\sigma^*, \sigma_{k+i}), \quad i = 1 \ldots N_p. \qquad (4.26d)$$

The result is an MILP problem which can be handled relatively easy via the construction of Chap. 3.

Illustrative example for the corner cutting problem

With the same double integrator dynamics as the ones presented in Sect. 4.1 we apply the corner cutting construction and observe the improvements.

The simulation results using the shadow region exclusion constraints from Corollary 4.2 are the trajectories (dashed lines with circle marker) depicted in Fig. 4.6. Note that the constraints apply only to the position component of the state. For comparison,

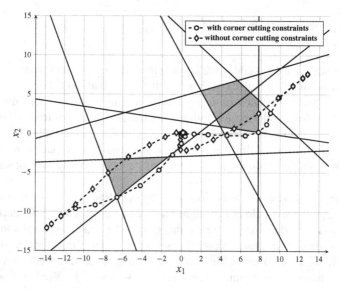

Fig. 4.6 Illustration of corner cutting simulation

we simulate from the same starting points without the corner cutting constraints (i.e., (4.26d)) and impose only obstacle avoidance constraints (dashed lines with diamond markers). As expected, we observe that the addition of corner cutting constraints leads to trajectories which avoid intra-sample obstacle collision.

4.3 Reference Governor Mechanism for Guaranteed Fault Detection

Nowadays the use of redundant components in applications is becoming increasingly common. Consequently, and with the present strict requirements on stability and performance, malfunctions in actuators, sensors or other systems components have become unacceptable [33]. As a consequence, a great deal of effort has been put into developing closed-loop systems which can tolerate faults while maintaining desirable performance and stability properties [11].

The occurrence of a fault imposes modifications not only in the feedback controller (which is to be expected for stability reasons) but also imposes the use of a reference governor and feedforward controller pair in order to deal with actuator degradation or to adjust the control input as required by performance or safety demands. Mixed-integer programming has been identified as a significant bottleneck in the optimization problems associated to the reference governor/MPC spanned by fault tolerant schemes.

In what follows we consider the multi-sensor control scheme illustrated in Fig. 4.7. The plant P, characterized by linear time-invariant dynamics

$$x^+ = Ax + Bu + Ew, \tag{4.27}$$

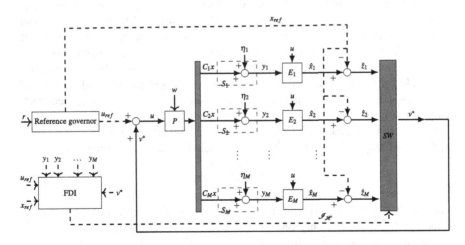

Fig. 4.7 Fault tolerant control multisensor scheme with reference governor

has to follow a reference

$$x_{ref}^+ = A x_{ref} + B u_{ref},\tag{4.28}$$

such that the plant tracking error $z = x - x_{ref}$ is minimized:

$$z^+ = x - x_{ref} = A z + B \underbrace{\left(u - u_{ref}\right)}_{v} + E w.\tag{4.29}$$

To each sensor $S_i, i \in \{1 \ldots M\}$, with output

$$y_i = C_i x_{ref} + \eta_i,\tag{4.30}$$

we attach an observer E_i which provides a state estimation:

$$\hat{x}_i^+ = A \hat{x}_i + B u + L_i \left(y_i - C_i \hat{x}_i\right),\tag{4.31}$$

and choose L_i in order to minimize the state estimation error $\tilde{x}_i = x - \hat{x}_i$:

$$\tilde{x}_i^+ = (A - L_i C_i) \tilde{x}_i + E w - L_i \eta_i.\tag{4.32}$$

The process noise w and measurement noises η_i are bounded: $w \in W$, $\eta_i \in N_i$.

We assume a simple fault event, i.e., the total output failure of a sensor output:

$$y_i = C_i x + \eta_i \xrightarrow[RECOVERY]{FAULT} y_i = 0 \cdot x + \eta_i^F,\tag{4.33}$$

where the measurement noise under fault, η_i^F, is bounded: $\eta_i^F \in N_i^F$.

In the resulting FTC scheme, the two elements which may vary are the design of the control action and of the residual signal which allows FDI. Various combinations are possible, each with its own strengths and weaknesses. The main idea is that the FDI mechanism has to detect instantaneously (or at least in a finite number of steps) the faulty sensor and send the information to the switching mechanism SW. Subsequently, the estimation provided by the faulty sensor is discarded from the ones admissible for feedback design and hence the overall performance and stability of the scheme are guaranteed (since only healthy sensors participate in the feedback design phase).

Analytically, the goal is to construct sets which bound the residual signal under healthy and faulty functioning (R_i^H and R_i^F) such that they do not intersect:

$$R_i^H \cap R_i^F = \emptyset.\tag{4.34}$$

In this case we can always ascertain the i-th sensor's functioning because the inclusion of the residual signal in only one of the residuals sets is unambiguous ($r_i \in R_i^H$ signifies healthy functioning, whereas $r_i \in R_i^F$ signifies faulty functioning).

Since these residual sets are parametrized after the reference values, the idea is to characterize the feasible domain in which these reference values can be taken such that the FDI condition is verified at all times. This domain is consequently used in a reference governor construction in order to provide feasible reference signals. The invariant sets, residual sets and particularities of the feasible domain strongly depend on the residual construction and feedback law design. In the rest of the section we will study several constructions and highlight their strong and weak points.

4.3.1 Output-Based Residual Case

Conceptually this represents the simplest construction. It assumes that the i-th residual (associated to the i-th sensor) is given as

$$r_i = y_i - C_i x_{ref},$$ (4.35)

composed from measurable quantities associated to the i-th sensor. The following expressions are obtained for the healthy and faulty functioning, respectively:

$$r_i^H = C_i z + \eta_i, \quad r_i^F = -C_i x_{ref} + \eta_i^F.$$ (4.36)

Using (4.36) together with the process and measurement noise bounds we can express the necessary and sufficient condition for exact FDI for a fault associated to sensor S_i as

$$(\{C_i z\} \oplus N_i) \cap (\{-C_i x_{ref}\} \oplus N_i^F) = \emptyset.$$ (4.37)

Assuming a static feedback law (keep in mind that $\hat{z}_l = \hat{x}_l - x_{ref} = x - x_{ref} + \tilde{x}_l = z + \tilde{x}_l$)

$$z = u_{ref} - K\hat{z}_l,$$ (4.38)

the tracking error dynamics become

$$z^+ = (A - BK)z - BK\tilde{x}_l + Ew.$$ (4.39)

Constructing the invariant set for (4.32), $\tilde{S}_i$, and introducing it in (4.39) leads to an invariant set S_z characterizing z, the tracking error. Consequently, the feasible reference domain for which (4.37) are valid, is defined by

$$D_{x_{ref}} \triangleq \{x_{ref} : (x_{ref}, S_z) \text{ verifies (4.37)}, \forall i = 1 \dots M\} = \mathbb{R}^n \setminus \bigcup_{i=1\dots M} P_i \quad (4.40)$$

with $P_i = \{x_{ref} : (\{C_i S_z\} \oplus N_i) \cap (\{-C_i x_{ref}\} \oplus N_i^F) \neq \emptyset\}$.
Using (4.40) we provide a reference governor construction:

$$u_{ref[0,\tau-1]}^* = \underset{u_{ref[0,\tau-1]}}{\arg\min} \left\{ \sum_{i=0}^{\tau-1} \left(||r_{[i]} - x_{ref[i]}||_{Q_r} + ||u_{ref[i]}||_{R_r} \right) + ||r_{[\tau]} - x_{ref[\tau]}||_{P_r} \right\}$$

$$(4.41)$$

subject to:

$$\begin{aligned} x_{ref[i]}^+ &= Ax_{ref[i]} + Bu_{ref[i]} \\ x_{ref[i]}^+ &\in D_{x_{ref}} \end{aligned}, \quad i = 0 \ldots \tau - 1. \quad (4.42)$$

where $r \in \mathbb{R}^n$ is the ideal reference to be followed, τ is the prediction horizon, and $Q_r \in \mathbb{R}^{n \times n}$, $P_r \in \mathbb{R}^{n \times n}$ and $R_r \in \mathbb{R}^{m \times m}$ are weighting matrices. The feedforward control action is then set to $u_{ref} = u_{ref[0]}^*$ which is the first component in the optimal sequence. Then, the optimization is reiterated by receding the reference window.

Remark 4.9 The converse is true. Assuming $x_{ref} \in X_{ref}$ bounded, a feasible domain for the tracking error can be deduced:

$$\begin{aligned} D_z &\triangleq \left\{ z : (z, X_{ref}) \text{ verifies } (4.37), \forall i = 1 \ldots M \right\} \\ &= \left\{ z : (\{C_i z\} \oplus N_i) \cap (\{-C_i X_{ref}\} \oplus N_i^F) = \emptyset, \ i = 1 \ldots N \right\}. \end{aligned} \quad (4.43)$$

◆

Remark 4.10 Note that $\underset{i=1\ldots M}{\bigcup} P_i$, the complement of $D_{x_{ref}}$ from (4.40), denotes the region in which *at least* a sensor's FDI is untrustworthy. Then $\underset{i=1\ldots M}{\bigcap} P_i$ has a clear meaning in the FDI context: any $x_{ref} \in \underset{i=1\ldots M}{\bigcap} P_i$ corresponds to un-detectable faults for *all* the sensors. On the other hand, for any $x_{ref} \in \underset{i=1\ldots M}{\bigcup} P_i \setminus \underset{i=1\ldots M}{\bigcap} P_i$ there exists a non-empty subset of sensors for which FDI is trustworthy. In other words, the persistent excitation constraint $x_{ref[i]}^+ \in D_{x_{ref}}$ from (4.42), in conjunction with a temporary elimination from the pool of available sensors of the untrustworthy ones, can be changed to $x_{ref[i]}^+ \notin \underset{i=1\ldots M}{\bigcap} P_i$. The same remarks hold for D_z from (4.43). ◆

Illustrative example for the output-based residual

Consider the linear time invariant system

$$x^+ = \begin{bmatrix} 1 & 0.1 \\ 0 & 1 \end{bmatrix} x + \begin{bmatrix} 0 \\ 0.5 \end{bmatrix} u + \begin{bmatrix} 0 \\ 0.1 \end{bmatrix} w, \quad (4.44)$$

which models the dynamics of a double integrator affected by a bounded noise $w \in W = \{w : -0.1 \le w \le 0.1\}$ and controlled through the signal u. The state is measured by a collection of sensors, defined by their output matrices and bounds upon their measurement noises under healthy and faulty functioning:

$$C_1 = \begin{bmatrix} 0.35 & 0.25 \end{bmatrix}, |\eta_1| \leq 0.1, |\eta_1^F| \leq 1$$
$$C_2 = \begin{bmatrix} 0.30 & 0.80 \end{bmatrix}, |\eta_2| \leq 0.1, |\eta_2^F| \leq 1$$
$$C_3 = \begin{bmatrix} 0.15 & 0.75 \end{bmatrix}, |\eta_3| \leq 0.1, |\eta_3^F| \leq 1. \tag{4.45}$$

The gain matrices L_i are chosen such that the estimator poles are placed in the interval $[0.75, 0.90]$ and the static gain matrix is taken as $K = \begin{bmatrix} 0.5 & 0.7 \end{bmatrix}$.

Based on these constructions and along the lines of [13] the invariant sets for the estimation errors (4.32) and tracking error (4.39) are computed and illustrated in Fig. 4.8a, b. Applying the FDI separation condition (4.37) we obtain the feasible domain (4.40). In Fig. 4.9, the gray area represents the region where no sensor is trustworthy $\left(\bigcap_{i=1...M} P_i \right)$, the dashed area represents the region where at least a sensor is untrustworthy $\left(\bigcup_{i=1...M} P_i \right)$ and the remaining represents the region where everything is well ($D_{x_{ref}}$).

What remains is to solve the constrained optimization problem (4.41)–(4.42). In either the original form or the variation shown in Remark 4.10 the resulting problem is mixed-integer as the feasible domain is non-convex (the complement of a union or intersection of degenerate polyhedra).

In Fig. 4.10 we depict an ideal reference trajectory (filled square markers) plotted against $\bigcap_{i=1...M} P_i$. Applying the mixed integer construction from the previous chapters we obtain a feasible reference trajectory (filled round markers). As each of the points of the feasible trajectory respects $x_{ref} \notin \bigcap_{i=1...M} P_i$ it follows that at least a sensor has a trustworthy FDI condition. This is illustrated in the right-hand side of the figure where the residual sets corresponding to a value $x_{ref} = \begin{bmatrix} 2.1 & 1.9 \end{bmatrix}^\top$ are depicted (represented in a 2D plot to better illustrate their (non) intersection, gray for the faulty residual sets and black for the healthy residual sets). As it can be seen, not all of the sensors

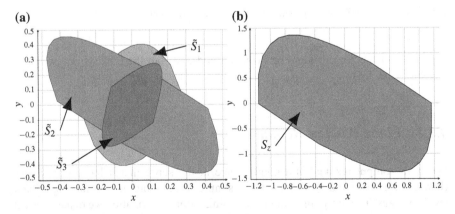

Fig. 4.8 Invariant sets for the FTC scheme. **a** Estimation error sets. **b** Tracking error set

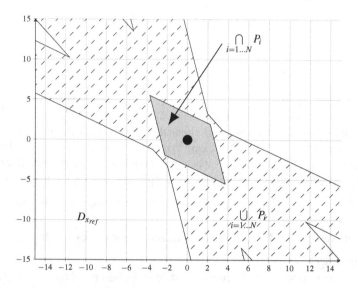

Fig. 4.9 Representation of $D_{x_{ref}}$ and its auxiliary sets

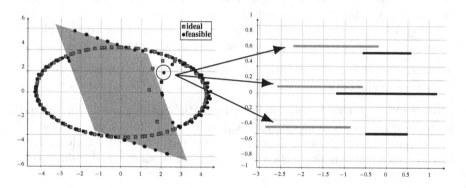

Fig. 4.10 Feasible reference computation and example of residual sets

have non-intersecting residual sets (the 1st and 2nd) but there exists one which is "usable" (the 3rd).

4.3.2 Finite Window Residual Case

While the output residual is a good choice in the sense that it has no dynamics (hence, no transitional behavior) and is simple to compute, it also has significant shortcomings. The most important is that using output information we make in effect a projection from the state space into the output space, i.e., we loose information.

In what follows we consider a "finite observation window" which uses information from the past $\tau + 1$ instants (interval $k \ldots k - \tau$) to define the residual:

$$
r_i = \Theta_{i,\tau}^+ \left(\begin{bmatrix} y_{i[-\tau]} \\ \vdots \\ y_{i,[0]} \end{bmatrix} - \begin{bmatrix} C_i x_{ref[-\tau]} \\ \vdots \\ C_i x_{ref,[0]} \end{bmatrix} - \begin{bmatrix} 0 & \ldots & 0 & 0 \\ C_i B & \ldots & 0 & 0 \\ & \ldots & \ldots & \\ C_i A^{\tau-1} B & \ldots & C_i A B & C_i B \end{bmatrix} \begin{bmatrix} v_{[-\tau]} \\ \vdots \\ v_{[-1]} \end{bmatrix} \right)
$$

$$
= \Theta_{i,\tau}^+ \left(y_{i[-\tau,0]} - \tilde{C}_{i,\tau} x_{ref[-\tau,0]} - \Gamma_{i,\tau} v_{[-\tau,-1]} \right)
$$

$$(4.46)$$

where

- τ represents the length of the residual horizon,
- $\Theta_{i,\tau} = \begin{bmatrix} C_i \\ C_i A \\ \ldots \\ C_i A^\tau \end{bmatrix}$ is the observation matrix; $\Theta_{i,\tau}^+$ is its left pseudo-inverse.

In other words, we have used all the available information ($y_{i[-\tau,0]}$, $x_{ref[-\tau,0]}$ and $v_{[-\tau,-1]}$) over the finite window interval to characterize a fault event starting at τ instants in the past.

Assuming that the system has been healthy (faulty) for the last τ instants of time, we have the healthy and faulty residual formulations:

$$
r_i^H = z_{[-\tau]} + \Theta_{i,\tau}^+ \left(\Phi_{i,\tau} w_{[-\tau,-1]} + \eta_{i[-\tau,0]} \right)
$$
$$
r_i^F = -x_{ref[-\tau]} - \Theta_{i,\tau}^+ \left(\Gamma_{i,\tau} \left(u_{ref[-\tau,-1]} + v_{[-\tau,-1]} \right) + \eta_{i[-\tau,0]}^F \right) \qquad (4.47)
$$

where $\Phi_{i,\tau} = \begin{bmatrix} 0 & \ldots & 0 & 0 \\ C_i E & \ldots & 0 & 0 \\ & \ldots & \ldots & \\ C_i A^{\tau-1} E & \ldots & C_i A E & C_i E \end{bmatrix}$.

Intuitively, we note that r_i^H stays around the origin and that r_i^F, due to the offset incorporated into $x_{ref,[-\tau]}$ and $u_{ref,[-\tau,-1]}$, will diverge sufficiently far away to obtain a clear separation between healthy and faulty behavior.

To obtain a feasible domain we need to consider bounds for the tracking error z and the feedback action v. This time (in contrast with the output residual construction), the FDI needs τ instants to take a decision (at current instant, it decides whether a fault has started τ time instants ago). Handling this decision delay can be tackled in two ways:

- the feedback is constructed with delayed information, i.e., we do not use information until it is not guaranteed to be healthy: $v = -K\hat{z}_{l,[-\tau]}$
- we use current information and accept the risk that it might be corrupted by a fault which started somewhere inbetween the current instant and $\tau - 1$ instants ago: $v = -K\hat{z}_l,$

In both cases the set bounding z changes, either because we have a closed-loop system with delay or because we use faulty information. The recomputation of the tracking error set and of the feasible domain is detailed below.

Using delayed information we have that

$$u = u_{\text{ref}} + K \hat{z}_{l[-\tau]} = u_{\text{ref}} + K \left(z_{[-\tau]} - \tilde{x}_{l[-\tau]} \right), \tag{4.48}$$

which leads to

$$z^+ = Az + BK z_{[-\tau]} + Ew - BK \tilde{x}_{l[-\tau]}. \tag{4.49}$$

Since this is a delay difference equation, an extended state model has to be considered in order to construct an invariant set:

$$z^+_{[-\tau,0]} = A_{z,\tau} z_{[-\tau,0]} + B_{z,l} \begin{bmatrix} \tilde{x}_{l[-\tau]} \\ w \end{bmatrix}, \tag{4.50}$$

where $A_{z,\tau}$ and $B_{z,\tau}$ are defined as

$$A_{z,\tau} = \begin{bmatrix} 0 & I & \dots & 0 & 0 \\ \dots & \dots & \dots & \dots & \dots \\ 0 & 0 & \dots & 0 & I \\ BK & 0 & \dots & 0 & A \end{bmatrix}, \quad B_{z,\tau} = \begin{bmatrix} 0 & 0 \\ \dots & \dots \\ 0 & 0 \\ -BK & E \end{bmatrix}.$$

It follows then that a bounding set, in which the tracking error, z, is guaranteed to reside as long as $z_{[-\tau,0]} \in S_{z_{[-\tau,0]}}$, can be defined:

$$S_z = \text{conv} \left\{ \bigcup_{j=-\tau,\dots,0} \text{proj}_{z_{[j]}} \left(S_{z_{[-\tau,0]}} \right) \right\}, \tag{4.51}$$

where the $\text{proj}_{z_{[j]}}$ operator denotes the projection of its argument along the given subspace $z_{[j]}$, i.e., $\text{proj}_{z_{[j]}} \left(S_{z_{[-\tau,0]}} \right) = \begin{bmatrix} 0 & \dots & 0 & I & 0 & \dots & 0 \end{bmatrix} S_{z_{[-\tau,0]}}$, with the identity matrix I located in the $j + \tau + 1$ position.

We can now characterize the feedback bounds

$$v_{[-\tau,0]} = \text{diag} \underbrace{(K, \dots, K)}_{\tau+1} \hat{z}_{l[-2\tau,-\tau]} = \text{diag} \underbrace{(K, \dots, K)}_{\tau+1} \left(z_{[-2\tau,-\tau]} - \tilde{x}_{l[-2\tau,-\tau]} \right).$$

$$\tag{4.52}$$

Using the previously computed sets we obtain a bound for the sequence $v_{[-\tau,0]}$:

$$v_{[-\tau,0]} \in \mathbb{V}, \quad \text{where } \mathbb{V} \triangleq \text{diag} \underbrace{(K, \dots, K)}_{\tau+1} \left[S_{z_{[-\tau,0]}} \oplus \bigcup_{i \in \mathbf{i}} \left(-(\tilde{S}_i)^\tau \right) \right]. \tag{4.53}$$

Using the above results we are now able to construct the FDI condition:

$$S_z \oplus \Theta_{i,\tau}^+ \left(\Phi_{i,\tau} W^{\tau+1} \oplus N_i^{\tau+1} \right)$$
$$\cap \left\{ -x_{\text{ref}[-\tau]} - \Theta_{i,\tau}^+ \Gamma_{i,\tau} u_{\text{ref}[-\tau,0]} \right\} \oplus \Theta_{i,\tau}^+ \left\{ -\Gamma_{i,\tau} \mathbb{V} \oplus \left(N_i^F \right)^{\tau+1} \right\} = \emptyset.$$
(4.54)

We can now redefine the reference feasible set:

$$\mathbb{D}_{\text{ref}} = \left\{ \left(x_{\text{ref}[-\tau]}, u_{\text{ref}[-\tau,-1]} \right) : (4.54) \text{ holds } \forall i \in \{1 \ldots M\} \right\},$$
(4.55)

which we use in the new reference governor construction explicited below:

$$u^* = \arg\min_{u_{\text{ref}[0,\sigma]}} \sum_{j=0}^{\sigma} \left(||r_{[j]} - x_{\text{ref}[j]}||_{Q_r} + ||u_{\text{ref}[j]}||_{R_r} \right),$$
(4.56)

subject to:

$$x_{\text{ref}[j]}^+ = A x_{\text{ref}[j]} + B u_{\text{ref}[j]},$$
$$\left(x_{\text{ref}[j-\tau]}, u_{\text{ref}[j-\tau,j]} \right) \in \mathbb{D}_{\text{ref}},$$
(4.57)

where $r \in \mathbb{R}^n$ is the ideal reference to be followed, $\sigma \geq \tau$ is the prediction horizon, and $Q_r \in \mathbb{R}^{n \times n}$ and $R_r \in \mathbb{R}^{m \times m}$ are weighting matrices. The current value of the input reference signal, $u_{\text{ref}}(k)$, is taken as the first element of the sequence u^*.

Remark 4.11 The use of feedback with potentially faulty information is another approach. As long as all the sensors are healthy the tracking error z and the estimation error $\tilde{x}_l$ lie inside their invariant sets, S_z and $\tilde{S}_i$ respectively. This no longer holds when one of the sensors has an incipient faulty behaviour and not enough time has elapsed for its detection. As a result, these sets need to be recomputed and after a fault detection a certain "deflation" time has to be ensured such that they reach back into the original invariant ones. ♦

Illustrative example for the finite window residual case

Using the same dynamics as before we consider now the finite window residual. We do not repeat all the constructions from before but rather emphasize some of the differences with respect to the output-based residual case.

Foremost is the influence of the window length (factor τ). In Fig. 4.11a different tracking error sets, corresponding to various τ values, are depicted. These are then used in Fig. 4.11b to illustrate the resulting feasible domains.[4] Note that the feasible

---

[4]This time the domain depends on both state and input references. For plotting purposes we assume the reference inputs bounded ($u_{ref} \in U_{ref}$) and construct the feasible domain as a function of x_{ref} only.

(a)   **(b)**

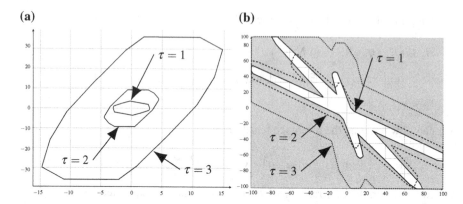

Fig. 4.11 Illustration of sets for multiple lengths of the residual window. **a** Tracking error set. **b** Feasible domain

domain is in all cases compact (and not a union of disjoint pieces as was the case for the output-based residual).

Further, while an open-loop analysis would assume that a longer τ leads to better FDI capabilities, the closed-loop behavior exhibited here shows that at some point the negative influence of the delayed feedback overcomes the residual improvement (intuitively this is shown through the degradation of the feasible domain at larger values of τ).

Lastly, Fig. 4.12 depicts the healthy, intermediate and faulty residual sets obtained for $\tau = 3$ and a particular x_{ref} value. As it can be seen, the first step under fault (nor

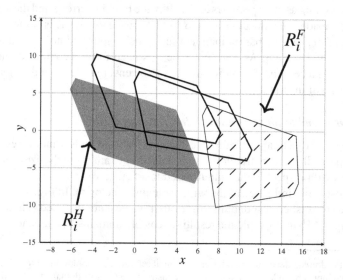

Fig. 4.12 Illustration of residual sets for the finite window construction

the second) does not suffice for set separation. Only after three consecutive steps under fault is the FDI unambiguous (the healthy residual set is not intersecting the three-steps-after-fault residual set).

4.4   Notes and Comments

We have shown in this chapter that various problems afflicting multi-agent systems can be treated unitary through the mixed-integer formalism. The issues discussed ranged from obstacle/agent collision avoidance and up to coverage conditions for a space partially occluded by obstacles. We have thus shown that the constructions presented in the previous chapter are able to provide characterizations of the various constraints affecting the agents. Moreover, the resulting constrained optimization problems are solved in reasonable time (thus making them amenable for real-time implementations).

Other applications using mixed-integer formulations (multi-agent formation control, trajectory tracking for a group of agent) are presented in [17]. Also, for larger number of agents Potential Field methods were used in combination with MPC highlighting the advantages and disadvantages of using MIP techniques for the collision avoidance issues [34]. Among others, Richards and How [5, 35], Earl and D'Andrea [36, 37] addressed multi-agent collision avoidance issues using classical mixed-integer formulations. Generally, we believe that the book of Mesbahi and Egerstedt [38] covers a good selection of issues in control of multi-agent systems. Also, the monographs of Grundel and Pardalos [7, 39] present a wide range of applications of cooperative systems in a control and optimization framework.

References

1. Wooldridge, M., Müller, J., Tambe, M.: Intelligent Agents II-Agent Theories, Architectures, and Languages, vol. 1037. Springer (1997)
2. Olfati-Saber, R., Murray, R.: Distributed cooperative control of multiple vehicle formations using structural potential functions. In: Proceedings of the 15th IFAC World Congress, pp. 346–352. Barcelona, Spain (2002)
3. Rimon, E., Koditschek, D.: Exact robot navigation using artificial potential functions. IEEE Trans. Robot. Autom. **8**(5), 501–518 (1992)
4. Richards, A., How, J.: Model predictive control of vehicle maneuvers with guaranteed completion time and robust feasibility. In: Proceedings of the 24th American Control Conference, vol. 5, pp. 4034–4040. Portland, Oregon, USA (2005)
5. Richards, A., How, J.: Aircraft trajectory planning with collision avoidance using mixed integer linear programming. In: IEEE (IEEE (ed.): Proceedings of the 21th American Control Conference. Anchorage, Alaska, USA (2002)), pp. 1936–1941
6. Shamma, J.: Cooperative Control of Distributed Multi-agent Systems. Wiley Online Library (2007)
7. Grundel, D., Murphey, R., Pardalos, P.: Cooperative Systems, Control and Optimization, vol. 588. Springer (2007)

8. Richalet, J., O'Donovan, D.: Predictive Functional Control: Principles and Industrial Applications. Springer (2009)
9. Rawlings, J., Mayne, D.: Postface to Model Predictive Control: Theory and Design (2011)
10. Camacho, E., Bordons, C.: Model Predictive Control. Springer (2004)
11. Zhang, Y., Jiang, J.: Bibliographical review on reconfigurable fault-tolerant control systems. Annu. Rev. Control 32(2), 229–252 (2008)
12. Seron, M., Zhuo, X.W., De Doná, J., Martínez, J.: Multisensor switching control strategy with fault tolerance guarantees. Automatica 44(1), 88–97 (2008)
13. Olaru, S., De Doná, J., Seron, M., Stoican, F.: Positive invariant sets for fault tolerant multisensor control schemes. Int. J. Control 83(12), 2622–2640 (2010)
14. Stoican, F., Olaru, S., Seron, M., De Doná, J.: Reference governor design for tracking problems with fault detection guarantees. J. Process Control 22(5), 829–836 (2012)
15. Prodan, I., Stoican, F., Olaru, S., Stoica, C., Niculescu, S.I.: Mixed-integer programming techniques in distributed MPC problems. In: Distributed MPC Made Easy, vol. 69, pp. 273–288. Springer (2013)
16. Prodan, I., Olaru, S., Stoica, C., Niculescu, S.I.: On the tight formation for multi-agent dynamical systems. In: KES—Agents and Multi-agent Systems—Technologies and Applications, pp. 554–565. Springer (2012)
17. Prodan, I.: Control of multi-agent dynamical systems in the presence of constraints. Ph.D. thesis, Supélec (2012), https://tel.archives-ouvertes.fr/tel-00783221/document
18. Althoff, M., Stursberg, O., Buss, M.: Computing reachable sets of hybrid systems using a combination of zonotopes and polytopes. Nonlinear Anal.: Hybrid Syst. 4(2), 233–249 (2010)
19. Grancharova, A., Grøtli, E.I., Ho, D.T., Johansen, T.A.: UAVs trajectory planning by distributed MPC under radio communication path loss constraints. J. Intell. Robot. Syst., pp. 1–20. Springer (2014)
20. Prodan, I., Bitsoris, G., Olaru, S., Stoica, C., Niculescu, S.: On the limit behavior for multi-agent dynamical systems. In: The IFAC Workshop on Navigation, Guidance and Control of Underwater Vehicles, pp. 106–111. Porto, Portugal (2012)
21. Stoican, F., Prodan, I., Olaru, S.: Hyperplane arrangements in mixed-integer programming techniques. Collision avoidance application with zonotopic sets. In: Proceedings of the IEEE European Control Conference, pp. 3155–3160 (2013)
22. Strutu, M.I., Stoican, T., Prodan, I., Popescu, D., Olaru, S.: A characterization of the relative positioning of mobile agents for full sensorial coverage in an augmented space with obstacles. In: Proceedings of the 21st Mediterranean Conference on Control and Automation, pp. 936–941. Platania-Chania, Crete, Grecee (2013)
23. Stoican, F., Grotli, E., Prodan, I., Oara, C.: On corner cutting in multi-obstacle avoidance problems. In: 5th IFAC Conference on Nonlinear Model Predictive Control, pp. 185–190. Seville, Spain (2015)
24. Murphy, R.R., Tadokoro, S., Nardi, D., Jacoff, A., Fiorini, P., Choset, H., Erkmen, A.M.: Search and rescue robotics. In: Springer Handbook of Robotics, pp. 1151–1173. Springer (2008)
25. Castillo-Rogez, J., Pavone, M., Nesnas, I., Hoffman, J.: Expected science return of spatially-extended in-situ exploration at small solar system bodies. In: IEEE Aerospace Conference, pp. 1–15 (2012)
26. Hubbell, N., Han, Q.: Dragon: detection and tracking of dynamic amorphous events in wireless sensor networks. IEEE Trans. Parallel Distrib. Syst. 23(7), 1193–1204 (2012)
27. Murray, R.M.: Recent research in cooperative control of multivehicle systems. Trans.-Am. Soc. Mech. Eng. J. Dyn. Syst. Meas. Control 129(5), 571 (2007)
28. Martini, H., Soltan, V.: Combinatorial problems on the illumination of convex bodies. Aequationes Math. 57(2), 121–152 (1999)
29. Lee, D., Lin, A.: Computational complexity of art gallery problems. IEEE Trans. Inf. Theory 32(2), 276–282 (1986)
30. Richards, A., Turnbull, O.: Inter-sample avoidance in trajectory optimizers using mixed-integer linear programming. Int. J. Robust Nonlinear Control 25, 521–526 (2015)

31. Maia, M.H., Galvão, R.K.H.: On the use of mixed-integer linear programming for predictive control with avoidance constraints. Int. J. Robust Nonlinear Control **19**, 822–828 (2009)
32. Deits, R., Tedrake, R.: Efficient mixed-integer planning for UAVs in cluttered environments. In: Proceedings of IEEE International Conference on Robotics and Automation, pp. 1–8 (2015)
33. Maciejowski, J., Jones, C.: MPC fault-tolerant flight control case study: flight 1862. In: Proceedings of the 4th IFAC Symposium on Fault Detection, Supervision and Safety of Technical Processes, pp. 121–126. Washington, DC, USA, (2003)
34. Prodan, I., Olaru, S., Stoica, S., Niculescu, S.I.: Predictive control for trajectory tracking and decentralized navigation of multi-agent formations. Int. J. Appl. Math. Comput. Sci. **23**(1), 91–102 (2013)
35. Richards, A., Bellingham, J., Tillerson, M., How, J.: Coordination and control of multiple UAVs. In: AIAA Guidance, Navigation, and Control Conference Monterey, pp. 1–11. CA (2002)
36. Earl, M., D'Andrea, R.: Modeling and control of a multi-agent system using mixed integer linear programming. In: Proceedings of the 40th IEEE Conference on Decision and Control, vol. 1, pp. 107–111. Orlando, Florida, USA (2001)
37. Fowler, J., D'Andrea, R.: Distributed control of close formation flight. In: In the Proceedings of the 41st IEEE Conference on Decision and Control, vol. 3, pp. 2972–2977. IEEE (2002)
38. Mesbahi, M., Egerstedt, M.: Graph Theoretic Methods in Multiagent Networks. Princeton University Press (2010)
39. Grundel, D., Pardalos, P.: Theory and Algorithms for Cooperative Systems, vol. 4. World Scientific Publishing Co Inc. (2004)

Chapter 5
Conclusions

The work summarized in this book develops and brings to light new insights in the use of mixed-integer formulations to efficiently describe non-convex and non-connected regions appearing in a wide range of applications in control theory. Elements like hyperplane arrangements, polyhedral and zonotopic sets, cell merging and Boolean algebra (see also Fig. 5.1 for a general view of the approach) were effectively merged together and have provided us with useful tools that were further integrated in a mixed-integer formalism. Mainly, these constructions have allowed us to reduce the number of binary variables appearing in the mixed-integer optimization problem. Once the overall problem was brought to an improved formulation specialized solvers have been employed.

The main characteristic and the numerical advantages of the proposed approach were illustrated throughout the manuscript by proof of concept examples with the hope that some ideas can be fixed via 2D, 3D illustration, while the theoretical results hold for any finite dimensional space. Furthermore, we have applied the obtained theoretical results over challenging applications in multi-agent control and fault tolerant schemes. Particular emphasis was given to obstacles and collision avoidance, coverage and corner cutting conditions, and reference governor design.

In general, we believe that the original results and insights in the description of a non-convex feasible region are useful not only for the multi-agent topics but also for other different control settings. In our opinion, some recent open research areas can benefit from the tools we have provided in this book and even by the entire mixed-integer formulation approach. For example, in microgrid systems the electrical power storage devices can be modeled using mixed-integer formulations [1, 2], in unit commitment problems for addressing an efficient schedule for load, renewable energy generation and demand response [3–6], for the design of optimization tasks to generate an efficient cable plan for substation automation systems [7], in data-driven robust optimization for formulating generalized disjunctive programs [8–10] or in chance-constrained programming [11, 12].

© The Author(s) 2016
I. Prodan et al., *Mixed-Integer Representations in Control Design*,
SpringerBriefs in Control, Automation and Robotics,
DOI 10.1007/978-3-319-26995-5_5

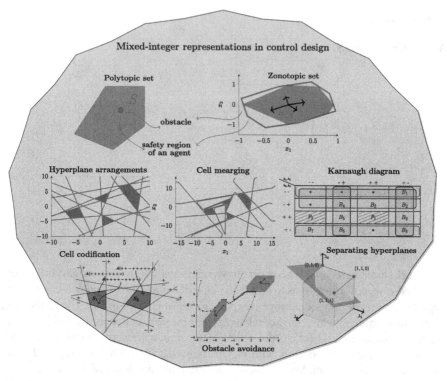

Fig. 5.1 Several representative tools used in the proposed mixed-integer formalism

Throughout the book we have showed that mixed-integer formulations provide one of the best ways of dealing with optimization problems with conflicting objectives. With all the valuable improvements that we carried out, the computational complexity is still highly dependent on the mixed-integer formulation and limits its usefulness to relatively small size problems. We believe that we can go further in advancing the novel geometrical interpretation of the non-convex constraints originated from the multi-agent problem by concentrating on more compact ways of describing the feasible region and exploiting its underlying combinatorial structure.

As a final conclusion line, the control and decision stage should at any mean try to model and formulate the problems to be solved in continuous domains and exploit the compactness and convexity whenever this is possible. However, if the problems are inherently non-convex and they involve alternative choices, then mixed-integer representations should not be dismissed, but considered with precaution and exploited according to the structural properties towards compact formulations.

References

1. Prodan, I., Zio, E.: A model predictive control for reliable microgrid energy management. Int. J. Electr. Power Energy Syst. **61**, 399–409 (2014)
2. Prodan, I., Zio, E., Stoican, F.: Fault tolerant predictive control design for reliable microgrid energy management under uncertainties. Energy, **91**, 20–34. Elsevier (2015)
3. Jiang, R., Wang, J., Guan, Y.: Robust unit commitment with wind power and pumped storage hydro. IEEE Trans. Power Syst. **27**(2), 800–810 (2012)
4. Arroyo, J., Conejo, A.: A parallel repair genetic algorithm to solve the unit commitment problem. IEEE Trans. Power Syst. **17**(4), 1216–1224 (2002)
5. Alguacil, N., Motto, A., Conejo, A.: Transmission expansion planning: a mixed-integer LP approach. IEEE Trans. Power Syst. **18**(3), 1070–1077 (2003)
6. Conejo, A., Arroyo, J.M., Contreras, J., Villamor, F.: Self-scheduling of a hydro producer in a pool-based electricity market. IEEE Trans. Power Syst. **17**(4), 1265–1272 (2002)
7. Sivanthi, T., Poland, J.: Efficient planning of substation automation system cables. In: Integration of AI and OR Techniques in Constraint Programming for Combinatorial Optimization Problems, pp. 210–214. Springer (2011)
8. Sung, C., Maravelias, C.: An attainable region approach for production planning of multiproduct processes. AIChE J. **53**(5), 1298–1315 (2007)
9. Ruiz, J., Grossmann, I.: Using redundancy to strengthen the relaxation for the global optimization of MINLP problems. Comput. Chem. Eng. **35**(12), 2729–2740 (2011)
10. Mitra, S., Grossmann, I.E., Pinto, J.M., Arora, N.: Optimal production planning under time-sensitive electricity prices for continuous power-intensive processes. Comput. Chem. Eng. **38**, 171–184 (2012)
11. Küçükyavuz, S.: On mixing sets arising in chance-constrained programming. Math. Program. **132**(1–2), 31–56 (2012)
12. Margellos, K., Goulart, P., Lygeros, J.: On the road between robust optimization and the scenario approach for chance constrained optimization problems. IEEE Trans. Autom. Control **59**(8), 2258–2263 (2014)

Appendix
Numerical Data for Illustrative Examples

A.1 Hyperplane Arrangements in $\mathbb{R}^2$ with 4 Hyperplanes

The hyperplane arrangement used in the illustrative example of Sect. 2.1 and the numerical observations from Sect. 2.3.2 is given by[1]:

$$H = \begin{bmatrix} 0.673 & 0.740 \\ 0.857 & -0.514 \\ -0.476 & 0.879 \\ 0.000 & 1.000 \end{bmatrix}, \quad k = \begin{bmatrix} 0.336 \\ 0.000 \\ 0.183 \\ 2.000 \end{bmatrix}. \tag{A.1}$$

The perturbed hyperplane arrangement described in the illustrative example of Sect. 2.1 and also analyzed in the illustrative example for region counting in Sect. 2.1.1 is given by:

$$H = \begin{bmatrix} 0.673 & 0.740 \\ 0.857 & -0.514 \\ 0.600 & -0.800 \\ 0.000 & 1.000 \end{bmatrix}, \quad k = \begin{bmatrix} 0.336 \\ 0.000 \\ 1.000 \\ 2.000 \end{bmatrix}. \tag{A.2}$$

A.2 Hyperplane Arrangements in $\mathbb{R}^3$ with 4 Hyperplanes

The parametrized hyperplane arrangement lifted in $\mathbb{R}^3$ as in (2.10) and presented in the illustrative example for hyperplane parametrization of Sect. 2.1.2 is given by the following numerical data:

---

[1] Matrices H and k store row-wise the elements defining the current hyperplane arrangement as in (2.1).

© The Author(s) 2016
I. Prodan et al., *Mixed-Integer Representations in Control Design*,
SpringerBriefs in Control, Automation and Robotics,
DOI 10.1007/978-3-319-26995-5

$$H = \begin{bmatrix} 0.2184 & -0.3944 & -0.4092 \\ 0.4686 & 0.1110 & -0.2335 \\ 0.0313 & 0.2788 & -0.3463 \\ -0.1749 & -0.0765 & -0.2190 \end{bmatrix}, \quad k = \begin{bmatrix} 1 \\ 1 \\ 1 \\ 1 \end{bmatrix}. \tag{A.3}$$

A.3 Hyperplane Arrangements in $\mathbb{R}^2$ with 3 Hyperplanes

The following hyperplane arrangement is used for the numerical observations provided in Sects. 2.3.2 and 3.6:

$$H = \begin{bmatrix} -0.0250 & -0.9997 \\ -0.6190 & 0.7854 \\ 0.9983 & -0.0583 \end{bmatrix}, \quad k = \begin{bmatrix} 4.7798 \\ 2.2037 \\ 5.9584 \end{bmatrix}. \tag{A.4}$$

The collection of forbidden tuples (see the description provided in (2.19)) of the above hyperplane arrangement used in the numerical consideration of Sects. 2.3.2 and 3.6 is: $\Sigma^{\bullet,1} = \{(+++)\}$, $\Sigma^{\bullet,2} = \{(+-+), (+--)\}$.

A.4 Hyperplane Arrangement in $\mathbb{R}^2$ with 8 Hyperplanes

The following hyperplane arrangement is used first in the illustrative example of non-convex regions in Sect. 2.3 and continued throughout the book in the illustrative example of merging procedures Sect. 2.3.1, numerical considerations Sect. 2.3.2, illustrative example for tuple allocation Sect. 3.2, illustrative example for mixed-integer formulations for the complement of a union of convex sets Sect. 3.3, illustrative example for mixed-integer formulations for the union of feasible cells Sect. 3.4, illustrative example for mixed-integer formulations for the feasible region characterized directly through the arrangement Sect. 3.5 and the numerical considerations in Sect. 3.6:

$$H = \begin{bmatrix} 0.7061 & -0.7081 \\ 0.0532 & -0.9986 \\ -0.9584 & -0.2855 \\ 0.1784 & 0.9840 \\ 1.0000 & 0 \\ 0.7577 & 0.6526 \\ -0.3427 & 0.9394 \\ 0.9172 & 0.3984 \end{bmatrix}, \quad k = \begin{bmatrix} 1.1403 \\ 2.9327 \\ 8.6224 \\ 1.5864 \\ 7.7650 \\ 8.6885 \\ 4.4814 \\ 3.9327 \end{bmatrix}. \tag{A.5}$$

Furthermore, the associated collection of forbidden tuples (see the description provided in (2.19)) of the above hyperplane arrangement used in the numerical consideration of Sects. 2.3.2 and 3.6 is:

$\Sigma^{\bullet,1} = \{(+ - + + + + + +), (+ + + - + + + -), (- + + - + + + -)\}.$
$\Sigma^{\bullet,2} = \{(- + + + + + + +), (+ - + + + + + +), (+ + + - + + + -), (- + + - + + + -), (- + + - + + + +)\}.$

A.5 Hyperplane Arrangement in $\mathbb{R}^2$ with 10 Hyperplanes

The following hyperplane arrangement is used for the numerical observations provided in Sects. 2.3.2 and 3.6:

$$H = \begin{bmatrix} -0.0134 & -0.9999 \\ -0.7822 & -0.6231 \\ -0.9999 & 0.0173 \\ -0.6312 & -0.7756 \\ 0.9844 & 0.1760 \\ 0.8663 & -0.4995 \\ -0.1076 & 0.9942 \\ -0.8091 & 0.5876 \\ 0.1496 & 0.9887 \\ -0.8177 & 0.5756 \end{bmatrix}, \quad k = \begin{bmatrix} 9.8619 \\ 10.0495 \\ 9.2534 \\ 0.6254 \\ 7.9396 \\ 7.2362 \\ 5.3507 \\ 2.9736 \\ 8.6598 \\ 7.3975 \end{bmatrix}. \qquad (A.6)$$

The associated collection of forbidden tuples of the hyperplane arrangement in (A.6) is given by:

$\Sigma^{\bullet,1} = \{(+ - + - + + + + + +), (+ - + - + + + - + +), (+ + + + + + + - - + +), (+ + + + + - + + + +)\}.$
$\Sigma^{\bullet,2} = \{(+ + - - + + + - + +), (+ - - - + + + - + +), (+ + - - + + + - + -), (+ + + + + - + + + +), (+ + + + + + + + + +), (+ + + + + + - + + +), (+ + + + - - + + - +)\}.$

A.6 Hyperplane Arrangement in $\mathbb{R}^2$ with 15 Hyperplanes

The hyperplane arrangement with 15 hyperplanes used for the numerical observations in Sects. 2.3.2 and 3.6 is given by:

$$H = \begin{bmatrix} 0.2810 & -0.9597 \\ -0.9994 & 0.0332 \\ -0.2261 & -0.9741 \\ -0.8722 & 0.4892 \\ -0.9016 & -0.4325 \\ 0.5457 & -0.8380 \\ 0.9818 & -0.1900 \\ 0.4778 & 0.8785 \\ -0.4713 & 0.8820 \\ 0.8572 & 0.5150 \\ -0.4854 & -0.8743 \\ 0.0298 & -0.9996 \\ 0.9940 & -0.1091 \\ -0.1431 & -0.9897 \\ 0.6968 & -0.7173 \end{bmatrix}, \quad k = \begin{bmatrix} 3.6436 \\ 3.5455 \\ 0.1421 \\ 7.1289 \\ 9.1531 \\ 1.0230 \\ 8.1203 \\ 11.8156 \\ 4.9609 \\ 6.1583 \\ 5.4883 \\ 9.1173 \\ 9.6930 \\ 2.8675 \\ 6.5968 \end{bmatrix}. \tag{A.7}$$

The associated collection of forbidden tuples of the above hyperplane arrangement is:

$\Sigma^{\bullet,1} = \{(+--++++++++-++-+), (+--+++++++++-++++), (+--++++++++++++++), (+--+++++++++++++-+), (+--++-++++-++-+), (+--++-+++++++-+), (-+-++-++++++++--), (-+-++--+++++++--), (++++++++++-+++++)\}$.

$\Sigma^{\bullet,2} = \{(+--++++++++-++-+), (+--++++++++++++-+), (+--++++++++++++), (+--+++++++-++++), (+--++-++++++++-+), (+--++-++++-++-+), (-+-++-+++++++++--), (-+-++--+++++++--), (-+-++-++++-++--), (-+-++-+++++++++-), (-+-++--+++++++-), (-++++--++++-++-+++++-), (-+-++--++-++++-), (-++++--+++++++-), (-++++-+++++++++-), (++++++++-+++++++), (+++-++++-+++++++), (+++-+++--+++++++), (+++++--++-++-++), (+++++---+-++-++), (++++++-++-++-++), (++++++-++-+++++++), (+++++++++-+++++++)\}$.

A.7 Hyperplane Arrangement in $\mathbb{R}^2$ with 20 Hyperplanes

The hyperplane arrangement with 20 hyperplanes used for the numerical observations in Sects. 2.3.2 and 3.6 is:

$$
H = \begin{bmatrix}
0.6788 & 0.7344 \\
-0.3861 & -0.9225 \\
0.0890 & -0.9960 \\
0.2177 & 0.9760 \\
-0.4780 & 0.8784 \\
0.9988 & -0.0480 \\
0.5009 & -0.8655 \\
0.8920 & 0.4520 \\
-0.8996 & 0.4368 \\
0.8263 & -0.5632 \\
-0.5066 & 0.8622 \\
-0.3709 & -0.9287 \\
0.7230 & 0.6909 \\
-0.9945 & -0.1046 \\
0.1375 & -0.9905 \\
-0.9825 & 0.1865 \\
0.3033 & 0.9529 \\
0.8982 & -0.4396 \\
0.9672 & 0.2540 \\
0.2379 & 0.9713
\end{bmatrix}, \quad
k = \begin{bmatrix}
0.1465 \\
4.4491 \\
0.6655 \\
2.0252 \\
7.1799 \\
0.5265 \\
7.4761 \\
6.2232 \\
1.4269 \\
0.0590 \\
4.5748 \\
1.8940 \\
2.6388 \\
3.3731 \\
7.5032 \\
7.8669 \\
6.4685 \\
6.2249 \\
0.3415 \\
8.5629
\end{bmatrix}. \tag{A.8}
$$

The collection of forbidden tuples of the hyperplane arrangement in (A.8) is:

$\Sigma^{\bullet,1} = \{(+++++++++-++-+-+-+++++++), (++-+++++-++-+-$
$++++++), (+++++++++-++++-+++++++), (-++-+-+++-$
$++-+++++-+), (-++-+-+-+-++-++++++-+), (-++-+-$
$+++-+++++++++-+), (-++--+++-+-+-+++-+++), (-+$
$+--+++-+-+-+++++++), (-++--+++-+-++++++++++)\}.$

$\Sigma^{\bullet,2} = \{(+--++++++-++-+-+++++++), (+--+++++++++-$
$+-++++++), (+--+++++++-+-+-++++++), (-++-+-$
$+++-++-+++++-+), (-++-+-+++++++-++++++-+), (-+$
$+-+-++++++-+++-+-+), (-++-+++++++++-+++++$
$-+), (-++--+++-+-++-+++++++), (-++--+++-+-+$
$+-++-+++), (-++-+++++-+-++-+++++++), (++-++--+$
$+-+-++++++--+), (-+-++--++-+-+++++--+), (-+-+$
$+-+++-+-+++++--+), (++-++-+++-+-+++++--+)\}.$

A.8 Hyperplane Arrangement in $\mathbb{R}^2$ with 25 Hyperplanes

The hyperplane arrangement with 25 hyperplanes used for the numerical observations in Sects. 2.3.2 and 3.6 is the following:

$$
H = \begin{bmatrix}
-0.0369 & -0.9993 \\
-0.8121 & -0.5835 \\
-0.8955 & 0.4451 \\
-0.3497 & -0.9369 \\
0.6692 & -0.7431 \\
0.9721 & 0.2347 \\
0.7124 & -0.7018 \\
-0.6766 & 0.7363 \\
-0.0134 & 0.9999 \\
-0.9949 & -0.1012 \\
0.8083 & -0.5887 \\
-0.0244 & 0.9997 \\
0.9941 & -0.1085 \\
0.5977 & -0.8017 \\
0.1435 & 0.9896 \\
0.9485 & 0.3167 \\
-0.4223 & -0.9065 \\
-0.9800 & 0.1989 \\
-0.5730 & -0.8195 \\
-0.8655 & 0.5009 \\
-0.8093 & -0.5874 \\
0.8359 & -0.5489 \\
0.9626 & 0.2709 \\
0.3205 & -0.9472 \\
-0.9975 & -0.0702
\end{bmatrix}, \quad k = \begin{bmatrix}
10.1488 \\
9.8786 \\
4.3678 \\
6.2746 \\
12.9809 \\
7.5973 \\
9.8859 \\
10.6347 \\
9.7468 \\
8.5397 \\
2.2924 \\
9.4343 \\
6.4009 \\
3.2242 \\
3.9199 \\
4.9772 \\
4.8237 \\
4.7993 \\
1.7939 \\
7.3355 \\
2.5228 \\
5.4023 \\
1.8647 \\
1.3794 \\
0.0316
\end{bmatrix}. \tag{A.9}
$$

The collection of forbidden tuples of the above hyperplane arrangement is given by:
$\Sigma^{\bullet,1} = \{(+ + - + + + + - + + + - + + - + + - + - + + + + -), (+ + - + + + + - + + + + + + - + + - + - + + + + -), (+ + - + + + + - + + + + + + - + + - + - - + + - + - - + + + -), (+ + - + + + + + + + + + + + + + + - - + - + + + -), (+ + + + + + + + + + + + + + + + + + - - + - + + + -), (+ + - + + + + + + + - + + + + + + + - - + - + + + -), (+ - + - + + + + + + + + + + + + - - - + - + + - -), (+ - + - + + + + + - + + + + + + - - - + - + + - -), (+ - + - + + + + + + + + + - + + - - - + - + + - -), (+ - + - + + + + + + + + + + - + + - + - + - + + - -), (+ - + - + + + + + + + + + + + + + - + - + - + + - -), (+ - +), (+ + + + + + + + + + + + - + + + + + + + + + + + + + + + + - +), (+ + + + + + + + + + + + + + + + - + + + + - +), (+ + + + + + + + + + + + + - + + + + + + + - + + + + + + - + + +), (+ + + + + + + + + + + - + + - + + + + - + + + + - +), (+ + + + + + + + + + + + - + + - + + - + + + + + + + + + - +), (+ + + + + + + + + + + + + - + + - + + - + + - + - + + +), (+ + + + + + - + + + - + - - + + + + - + + - - - +), (+ + + + + + - + + + - + - - + + + + + + + - - - +), (+ + + + + + - + + + - + - - + - - + + - + - - + - +), (+ + + + + + - + + + - + - - + - - + - - + - + + - - - +), (+ + + + + + - + + + - + - - + - - + - + + - - - +), (+ + + + + + - + + + - + - - + - - + - + +$

$+++++++-++++-+++++++-++)$, $(+++++++++++-+-++-$
$+++++++-++)$, $(+++++-++++-+-++-+++++++-++)$, $(++$
$+++-+++++-+-++-++++++--++)$, $(+++++++++++-+++$
$+-+++++++--+)$, $(+++++-+++++++-+--++++++-++)\}$.
$\Sigma^{\bullet,2} = \{(++-++++-+++-++-++-+-++++-)$, $(++-+++++-$
$++++++-++-+-++++-)$, $(++-+++++-+++++++-++-+--+$
$++-)$, $(++-+++++++++++++++--+-+++-)$, $(++++++++$
$+++++++++-+-+++-)$, $(++-+++++++-+++++++++--+-+$
$++-)$, $(+-+-++++++++++++++---+-++--)$, $(+-+-++++$
$+-+++++++---+-++--)$, $(+-+-+++++++++-++--+-+$
$+--)$, $(+-+-+++++++++++-++-+-+-++--)$, $(+-+-++++$
$+++++++++-+-+-++--)$, $(++++++++++++++++++++++++$
$+-+)$, $(++++++++++++-+++++++-++++-+)$, $(++++++++$
$+++++++++++-++++-+)$, $(++++++++++++-+++++++++++$
$+-+)$, $(++++++++++++-++-++++-++++-+)$, $(++++++++$
$++-++-+++++++++-+)$, $(++++++++++++-++-++-+-+++$
$+-+)$, $(++++++-+++-+--++-+-++---+)$, $(++++++-+$
$++-+--++++-+++-++---+)$, $(++++++-+++-+--+-+++++-$
$--+)$, $(++++++-+++-+--++++++++---+)$, $(++++++-+$
$++-+--+-++-++---+)$, $(++++++-+++-+--+--+-++-$
$--+)$, $(+++++++++++-++++-+++++++-++)$, $(++++++++$
$+++++++-++++++-++)$, $(+++++-++++-+-++-++++++$
$-++)$, $(+++++++++++-+-++-++++++-++)$, $(++++++++$
$++-+++++-+++++++--+)$, $(+++++-++++-+-++-++++$
$+--++)$, $(+++++-+++++++-+--++++++-++)$, $(++-+++$
$++-++-++-++-+++++++-++)$, $(++-++++++-++-++-+++$
$++++-+-)$, $(++-+++++-++-++-++++-++-+-)$, $(++-+$
$++++-++-++-++++-++-++)$, $(++-+++++--++-++-+$
$+++-++-+-)$, $(++-+++++--++-++-++++-++-++)$, $(++$
$-+++++--++-++-+++++++-++)$, $(+--+++++++-++++$
$++-----+++-)$, $(+--+++++++-+++++++----+++-)\}$.

A.9   Hyperplane Arrangement in $\mathbb{R}^2$ with 30 Hyperplanes

The hyperplane arrangement with 30 hyperplanes used for the numerical observations
in Sects. 2.3.2 and 3.6 is given by:

$$
H = \begin{bmatrix}
-0.1069 & -0.9943 \\
0.6363 & -0.7714 \\
-0.8853 & -0.4651 \\
0.0791 & -0.9969 \\
-0.9957 & 0.0931 \\
0.1035 & 0.9946 \\
-0.8545 & 0.5195 \\
-0.8676 & -0.4972 \\
-0.1081 & 0.9941 \\
-0.9871 & -0.1601 \\
0.7955 & 0.6060 \\
-0.0403 & 0.9992 \\
0.8971 & -0.4418 \\
0.0675 & 0.9977 \\
-0.6650 & 0.7468 \\
0.0295 & -0.9996 \\
0.9414 & -0.3372 \\
0.9547 & 0.2974 \\
0.4854 & -0.8743 \\
0.8924 & 0.4512 \\
-0.0528 & -0.9986 \\
0.7739 & -0.6333 \\
0.9537 & 0.3006 \\
0.2968 & -0.9549 \\
0.9570 & 0.2901 \\
0.0816 & -0.9967 \\
0.9487 & -0.3163 \\
0.3705 & 0.9288 \\
-0.6812 & 0.7321 \\
-0.8789 & -0.4771
\end{bmatrix}, \quad
k = \begin{bmatrix}
10.0882 \\
2.7396 \\
8.2881 \\
4.8767 \\
8.9032 \\
4.2338 \\
7.1722 \\
0.2417 \\
10.0049 \\
7.3457 \\
6.1603 \\
5.0181 \\
3.6768 \\
9.5919 \\
3.5357 \\
3.3916 \\
9.4935 \\
9.2727 \\
2.7050 \\
4.7929 \\
9.1356 \\
11.2074 \\
5.1821 \\
7.5297 \\
0.5360 \\
1.7437 \\
2.1479 \\
2.3465 \\
2.3203 \\
1.8710
\end{bmatrix}.
\tag{A.10}
$$

The associated collection of forbidden tuples of the above hyperplane arrangement is the following:
$\Sigma^{\bullet,1} = \{(++--+++-+-+++++-++++++++++-+++-), (++---$
$+++-+-+++++-++-+++++++-+++-), (++--+++-+++++$
$+-++-+++++++-+++-), (++--+++-+++++++-++++++++$
$+-+++-), (++--+++-+-+++++-++-+-++++-+++-), (+-$
$+-+++-++++-++-++-++++----+++), (+-+-+++++++$
$-++-++-++++----+++), (+-+-+++++++-++--+-+$
$+++----+++), (++++++++++++++++++++++++++++++$
$++), (++++++++++++++++++++++++++-+++++), (++++++$
$+-+++++++++++++++++++++++), (+++++---+++-++-+$
$++++++++++--+), (+++++---+++-++-+++++++++++$
$+---), (++++----+++-++-+++++++++++++---), (++++$
$+---+++-++-+++++++++++++--), (+++++-+++++--++$

+++++−++−+−+−−−+), (+++++−++++−−+++++++−++−+
−+−−++), (+++++−++++−−+++++−+−++−+−+−−++), (++
+++−++++−−++++++−++−+−++−−+), (+++++−++++−−
+++++++−++++−++−−+), (+++++−++++−−++++++++−++
++−++−++), (+++++−++++−−+++++++−++−+−++−++)}.
$\Sigma^{\bullet,2} = \{$(++−−+++−+−+++++−++++++++−+++−), (++−−++
+−+−+++++−++−+++++++−+++−), (++−−+++−+++++++−
++−+++++++−+++−), (++−−+++−+−+++++−++−+−++++−
+++−), (++−−+++−++++++++−++++++++++−+++−), (+−+−
++++++++−++−++−++++−−−+++), (+−+−+++−++++−+
+−++−++++−−−+++), (+−+−++++++++−++−−+−++++−
−−−+++), (+++++++++++++++++++++++++++++++++), (++
+++++−+++++++++++++++++++++++), (+++++++++++
++++++++++++−+++++), (+−+++++++++−+++++−−++
−+−+−+++), (+−+++++++++++−+++++−−++−+−+−−++),
(+−+++++++++−+−+++++−−++−+−+−−++), (+−++++++
++−+−+++++−−++−+−−−−++), (+−+++++++++−+−+++
++−−++−+−−−+++), (+−+++++++++−+−+++++−−++−+
−+−+++), (+−++++++++++−+++++−−++−+−−−+++), (++
+++−++++−−+++++++−++−+−+−−++), (+++++−++++
−−+++++++−++−+−+−−−+), (+++++−++++−−++++++−
+−++−+−+−−++), (+++++−++++−−+++++++−++−+−+
+−−+), (+++++−++++−−++++++++−++++−++−−+), (++++
+−+++++−−+++++++−++−+−++−++), (+++++−++++−−
+++++++−++++−++−++), (+++++−−−+++−++−+++++
+++++++−−+), (+++++−−−+++−++−++++++++++++++−
−−), (++++−−−−+++−++−+++++++++++++−−−), (++++
+−−−+++−++−+++++++++++++−−), (+++++−++−++−
+−−+++++++++++−++−−+), (+++++−++−+−−+−−+++++
++++−++−−+), (+++++−++−+−−+−−+++++−++++−++−
−+), (++−+++++−+−+++++−+++++++++−+++−), (++−+++
+−+−+++++−+++++++−++−−), (++−++++−+−+−++++
++++++++++++−++−−), (++−++++−+−+++++−+++++++
++++++−−), (++−++++−+−++++−+++++++++++−++−−)}.

Series Editor's Biographies

Tamer Başar is with the University of Illinois at Urbana-Champaign, where he holds the academic positions of Swanlund Endowed Chair, Center for Advanced Study Professor of Electrical and Computer Engineering, Research Professor at the Coordinated Science Laboratory, and Research Professor at the Information Trust Institute. He received the B.S.E.E. degree from Robert College, Istanbul, and the M.S., M.Phil, and Ph.D. degrees from Yale University. He has published extensively in systems, control, communications, and dynamic games, and has current research interests that address fundamental issues in these areas along with applications such as formation in adversarial environments, network security, resilience in cyber-physical systems, and pricing in networks.

In addition to his editorial involvement with these Briefs, Basar is also the Editor-in-Chief of Automatica, Editor of two Birkhäuser Series on Systems & Control and Static & Dynamic Game Theory, the Managing Editor of the Annals of the International Society of Dynamic Games (ISDG), and member of editorial and advisory boards of several international journals in control, wireless networks, and applied mathematics. He has received several awards and recognitions over the years, among which are the Medal of Science of Turkey (1993); Bode Lecture Prize (2004) of IEEE CSS; Quazza Medal (2005) of IFAC; Bellman Control Heritage Award (2006) of AACC; and Isaacs Award (2010) of ISDG. He is a member of the US National Academy of Engineering, Fellow of IEEE and IFAC, Council Member of IFAC (2011–2014), a past president of CSS, the founding president of ISDG, and president of AACC (2010–2011).

Antonio Bicchi is Professor of Automatic Control and Robotics at the University of Pisa. He graduated from the University of Bologna in 1988 and was a postdoc scholar at M.I.T. A.I. Lab between 1988 and 1990. His main research interests are in:

- dynamics, kinematics and control of complex mechanical systems, including robots, autonomous vehicles, and automotive systems;
- haptics and dextrous manipulation; and theory and control of nonlinear systems, in particular hybrid (logic/dynamic, symbol/signal) systems.

© The Author(s) 2016   105
I. Prodan et al., *Mixed-Integer Representations in Control Design*,
SpringerBriefs in Control, Automation and Robotics,
DOI 10.1007/978-3-319-26995-5

- theory and control of nonlinear systems, in particular hybrid (logic/dynamic, symbol/signal) systems.

He has published more than 300 papers in international journals, books, and refereed conferences.

Professor Bicchi currently serves as the Director of the Interdepartmental Research Center "E. Piaggio" of the University of Pisa, and President of the Italian Association or Researchers in Automatic Control. He has served as Editor in Chief of the Conference Editorial Board for the IEEE Robotics and Automation Society (RAS), and as Vice President of IEEE RAS, Distinguished Lecturer, and Editor for several scientific journals including the *International Journal of Robotics Research, the IEEE Transactions on Robotics and Automation, and IEEE RAS Magazine*. He has organized and co-chaired the first WorldHaptics Conference (2005), and Hybrid Systems: Computation and Control (2007). He is the recipient of several best paper awards at various conferences, and of an Advanced Grant from the European Research Council. Antonio Bicchi has been an IEEE Fellow since 2005.

Miroslav Krstic holds the Daniel L. Alspach chair and is the founding director of the Cymer Center for Control Systems and Dynamics at University of California, San Diego. He is a recipient of the PECASE, NSF Career, and ONR Young Investigator Awards, as well as the Axelby and Schuck Paper Prizes. Professor Krstic was the first recipient of the UCSD Research Award in the area of engineering and has held the Russell Severance Springer Distinguished Visiting Professorship at UC Berkeley and the Harold W. Sorenson Distinguished Professorship at UCSD. He is a Fellow of IEEE and IFAC. Professor Krstic serves as Senior Editor for *Automatica and IEEE Transactions on Automatic Control* and as Editor for the Springer series *Communications and Control Engineering*. He has served as Vice President for Technical Activities of the IEEE Control Systems Society. Krstic has co-authored eight books on adaptive, nonlinear, and stochastic control, extremum seeking, control of PDE systems including turbulent flows and control of delay systems.

Index

© The Author(s) 2016
I. Prodan et al., *Mixed-Integer Representations in Control Design*,
SpringerBriefs in Control, Automation and Robotics,
DOI 10.1007/978-3-319-26995-5

Printed in the United States
By Bookmasters